P9-CCX-473

SMALL
ENGINES

Other Publications:

PLANET EARTH
COLLECTOR'S LIBRARY OF THE CIVIL WAR
LIBRARY OF HEALTH
CLASSICS OF THE OLD WEST
THE EPIC OF FLIGHT
THE GOOD COOK
THE SEAFARERS
THE ENCYCLOPEDIA OF COLLECTIBLES
THE GREAT CITIES
WORLD WAR II
THE WORLD'S WILD PLACES
THE TIME-LIFE LIBRARY OF BOATING
HUMAN BEHAVIOR
THE ART OF SEWING
THE OLD WEST
THE EMERGENCE OF MAN
THE AMERICAN WILDERNESS
THE TIME-LIFE ENCYCLOPEDIA OF GARDENING
LIFE LIBRARY OF PHOTOGRAPHY
THIS FABULOUS CENTURY
FOODS OF THE WORLD
TIME-LIFE LIBRARY OF AMERICA
TIME-LIFE LIBRARY OF ART
GREAT AGES OF MAN
LIFE SCIENCE LIBRARY
THE LIFE HISTORY OF THE UNITED STATES
TIME READING PROGRAM
LIFE NATURE LIBRARY
LIFE WORLD LIBRARY
FAMILY LIBRARY:
HOW THINGS WORK IN YOUR HOME
THE TIME-LIFE BOOK OF THE FAMILY CAR
THE TIME-LIFE FAMILY LEGAL GUIDE
THE TIME-LIFE BOOK OF FAMILY FINANCE

This volume is part of a series offering homeowners detailed instructions on repairs, construction and improvements they can undertake themselves.

HOME REPAIR
AND IMPROVEMENT

SMALL ENGINES

BY THE EDITORS OF
TIME-LIFE BOOKS

TIME-LIFE BOOKS
ALEXANDRIA, VIRGINIA

THE CONSULTANTS: John Lawrence teaches auto mechanics and small-engine repair at Thomas A. Edison High School in Alexandria, Virginia, and at Northern Virginia Community College. He has been a quality-control inspector of small engines for a Japanese manufacturing firm and has been involved with all aspects of small-engine repair since 1962.

Roswell W. Ard is a consulting engineer and a professional home inspector in northern Michigan. He has written professionally on construction techniques.

Harris Mitchell, special consultant for Canada, has worked in the field of home repair and improvement since 1950. He writes a syndicated newspaper column, "You Wanted to Know," and is author of a number of books on home improvement.

For information about any Time-Life book, please write:
Reader Information
Time-Life Books
541 North Fairbanks Court
Chicago, Illinois 60611

Library of Congress Cataloguing in Publication Data
Main entry under title:
Small engines.
(Home repair and improvement; 35)
Includes index.
1. Internal combustion engines, Spark ignition.
I. Time-Life Books. II. Series.
TJ790.S58 1982 621.43'4 82-10304
ISBN 0-8094-3510-1 (retail ed.)
ISBN 0-8094-3511-X (lib. bdg.)
ISBN 0-8094-3512-8 (reg. bdg.)

First printing. Printed in U.S.A.
Published simultaneously in Canada.
School and library distribution by Silver Burdett Company, Morristown, New Jersey.

Contents

1 Learning to Work with Machines

A critical measurement. During an overhaul, a micrometer is used to check the dimensions of a cam, which opens and closes an engine valve with split-second timing. The measurement is critical—a hairbreadth of wear can throw the timing off and damage the engine—but the job is surprisingly simple. The mechanic turns the sleeve at top right until rods in the horseshoe-shaped frame touch the cam, then takes the reading from scales on the shaft and sleeve *(page 13)*. Although the micrometer is sensitive and accurate to a thousandth of an inch, a beginner can master its use in a few minutes.

Yard work, once the backbreaking bane of the homeowner, has been revolutionized by new breeds of outdoor power tools. Preparing the soil for a flower or kitchen garden used to take a week or more of hard labor with spade, hoe and pitchfork. But a modern, laborsaving power tiller can turn over the land for a garden measuring 100 by 150 feet in less than a day. If a dead tree disfigures the yard, the homeowner is far more likely to remove it with a chain saw rather than a slow, clumsy handsaw. And few lawns are still groomed with a push-and-pull manual mower; today, the all-but-universal tool of choice is a rotary power mower.

The driving force in all these tools is a one-cylinder, gasoline-powered engine. Manufacturers and engineers classify it as a small engine, and some models are very small indeed: A typical weed-trimmer engine weighs less than five pounds, fills about as much space as a cantaloupe and contains three major moving parts. On the other hand, the largest and heaviest one-cylinder engines put out enough power to drive a riding mower or a garden tractor. In all their diverse incarnations, small engines number nearly 70 million in the United States—roughly the number of black-and-white television sets. They are the most faithful of domestic servants; with proper maintenance and regular service, a well-made engine should last for approximately 10 years.

Maintaining and servicing small engines are the basic subjects of this book. The chapter that follows deals with the tools and equipment you will need for these tasks. Some items, such as hammers and screwdrivers, are part of every home workshop; others are highly specialized and even exotic. To remove a set of engine valves, for example, you should have a pincer-like device whose sole purpose is to squeeze the valve springs tight; to smooth the inner wall of a cylinder, the ideal tool is a special hone fitted with a pair of spinning abrasive stones. Such seldom-used tools can be expensive, but even the most expensive tools are money savers in the long run. The cost of the parts for tuning a small engine *(Chapter 2)* may amount to about one fifth the price of a repair-shop job. And as you gain familiarity with the engine, you will become more confident and skillful in tackling more difficult repair jobs. In effect, tuning the engine regularly to keep it running at its best is a training course for the time when a complete overhaul *(Chapter 3)* is needed.

Any engine, of course, is only as useful as the machine it drives. The final chapter, therefore, deals with such tools as saws and mowers, and with the gears and drive mechanisms that link an engine to the business end of a tool—the blades or other cutting parts that do the actual work.

A Tool Kit for Home Mechanics

In addition to such stand-bys as flat-tipped and Phillips screwdrivers and adjustable wrenches, small-engine work requires a few general-purpose mechanic's tools and some specialized ones for such jobs as tune-ups or engine overhauls. All of these tools are readily available at large hardware stores or at auto-supply stores. Although most small engines are American-made, you may run across some foreign-made engines, especially in chain saws and weed cutters; for these you may need metric tools.

□ The core of any mechanic's tool kit is a set of socket wrenches. A common set and some helpful accessories are shown here and demonstrated on pages 10-11. Two special wrench handles are also used with sockets. An impact driver frees stuck bolts; with an adapter for a screwdriver head, it can also loosen frozen screws. A torque wrench, with a calibrated gauge on the handle, accurately measures the amount of twisting force being applied to fasteners.

□ Pliers—both the locking-grip and long-nose types—are essential for holding workpieces securely. Adjustable snap-ring pliers, designed to remove snap rings *(page 14)*, are also useful.

□ Two striking tools are musts for engine work: a plastic-tipped mallet, for tapping parts without marring them, and a ball-peen hammer, for striking heavy tools.

□ Other general-purpose tools include a set of hex wrenches ranging in size from 1⁄16 to 3⁄8 inch, combination wrenches sized from 1⁄4 to 11⁄16 inch, a set of punches and a wire brush.

□ Specialized tools needed for a small-engine tune-up generally fall into two categories. Precision measuring instruments include various gauges and a machinist's rule; testing equipment includes a battery tester, a compression tester and a continuity tester. You also will need an ignition-points file *(page 43)* and a flywheel puller *(page 42)*.

□ Essential tools to have on hand for doing an engine overhaul—servicing the major parts—are a telescoping gauge, a micrometer, a piston-ring expander, a cylinder hone, a valve-lapping tool and a valve-spring compressor.

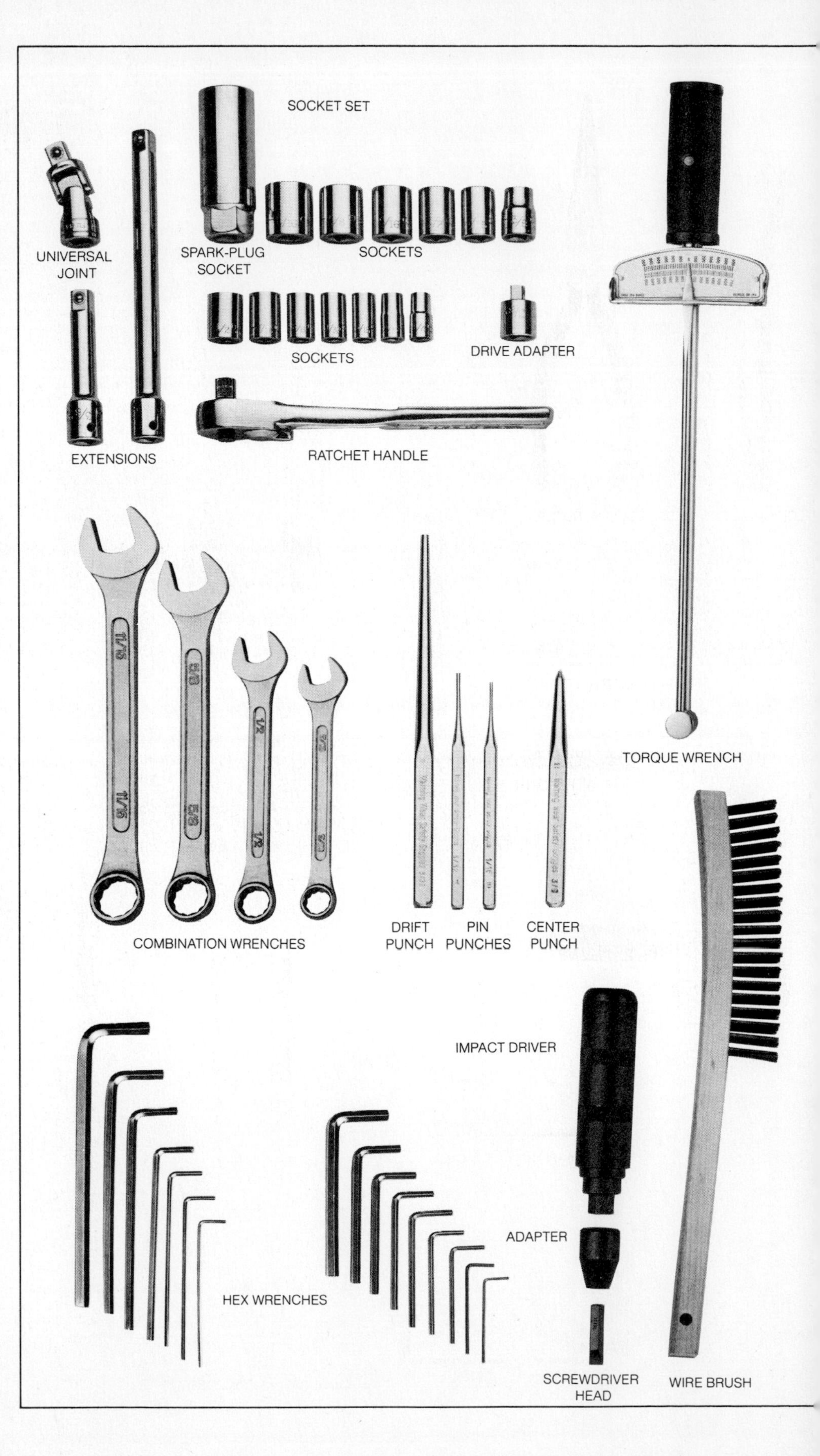

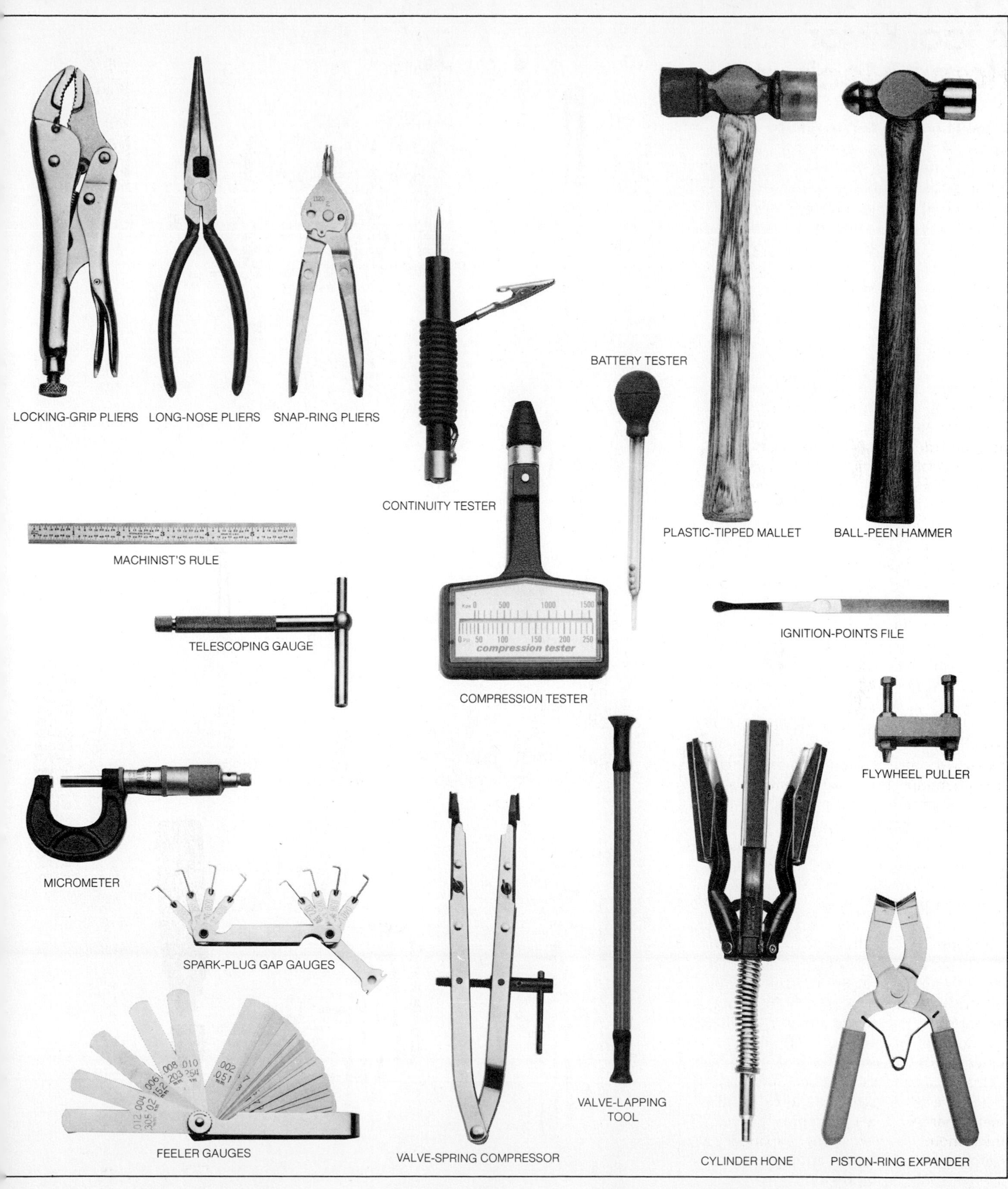
LOCKING-GRIP PLIERS
LONG-NOSE PLIERS
SNAP-RING PLIERS
CONTINUITY TESTER
BATTERY TESTER
PLASTIC-TIPPED MALLET
BALL-PEEN HAMMER
MACHINIST'S RULE
TELESCOPING GAUGE
compression tester
COMPRESSION TESTER
IGNITION-POINTS FILE
FLYWHEEL PULLER
MICROMETER
SPARK-PLUG GAP GAUGES
FEELER GAUGES
VALVE-SPRING COMPRESSOR
VALVE-LAPPING TOOL
CYLINDER HONE
PISTON-RING EXPANDER

Basic Wrenches for an Engine's Nuts and Bolts

The multitude of nuts and bolts on a small engine make work horses of two tools: the socket wrench—a ratchet handle fitted with a socket—and the torque wrench. Socket wrenches are used both to tighten and to loosen fasteners. Torque wrenches are used with sockets to turn a fastener with a measured amount of force so that parts are held together under a specified tension.

A typical socket set comes with a ratchet handle and a range of socket sizes. The head of the ratchet handle has a release gear that allows the handle to swing back while the socket stays in place. The action can be reversed, for either tightening or loosening, by flipping a control lever. The ratchet head has a square protrusion, the square drive, that fits into a matching square hole in one end of a socket. The handles come in a range of square-drive sizes—¼-, ⅜- and ½-inch. For work on small engines, the ⅜-inch drive size is most useful.

Many socket sets have an adapter so that a ratchet can be converted to a different square-drive size—for example, from ⅜ inch to ¼ inch or ½ inch. But adapters should not be used to convert a ½-inch drive to a ¼-inch, or vice versa: Reducing drive size this much strains the socket; increasing it strains the ratchet.

A socket set usually includes sockets graduated in 1⁄16-inch increments. Metric sets sized in millimeters are also made. The sockets have either 6 or 12 corners, or points, inside the socket recess. A 6-point socket is less likely to round the fastener head. A 12-point socket is useful in tight places because it need be turned only half as far to engage a nut.

Sockets are made in two lengths. Standard sockets have ½-inch-deep recesses; so-called deep-well sockets have 1½-inch-deep hollows for getting to nuts that are set too far down a shaft for a standard socket to reach. All spark-plug sockets are of this type.

A variety of socket accessories are made to reach awkwardly placed fasteners. Extensions ranging from 1 inch to 3 feet long are available; they fit onto a ratchet handle to provide extra reach. For small-engine repair work, however, extensions of 6 or 8 inches are adequate.

To get to fasteners that cannot be turned at a right angle, an accessory called the universal joint swivels as much as 45°, enabling the ratchet and socket to work at odd angles and around obstructions. A flexible extension that snaps onto the ratchet handle bends farther than a universal joint, even into a U shape. But it should be used only to turn bolts already loosened.

The torque wrench, which measures turning force, is a precision tool required in small-engine repair when an engine manual calls for an exact amount of twisting tension on a fastener. For example, proper torque is critical for bolts on the cylinder head and those that hold the connecting rod to the crankshaft.

The torque reading is the product of the length of the wrench handle and the applied force. It is expressed in inch-pounds or foot-pounds and, in the metric system, in newton-meters.

A variety of styles of torque wrenches are sold, including the type with a flexible arm and a needle gauge *(opposite, top)* and the more sophisticated, more expensive, click-type torque wrench *(opposite, bottom)*. The latter has the advantage of a ratchet head and an audible signal that tells you when the proper torque has been applied. This wrench is preset for the desired torque, and it resets itself after each fastener.

The proper way to use a torque wrench is to support the socket on the fastener with one hand and apply turning effort at right angles to the handle. For the most accurate reading, you should clean the fastener threads and apply lubricant to the threads and under the head of the bolt. Do not use the torque wrench on a fastener that is already tight; loosen it with a socket wrench before tightening it to the correct specification.

Socket-wrench Adapters for Hard-to-Reach Fasteners

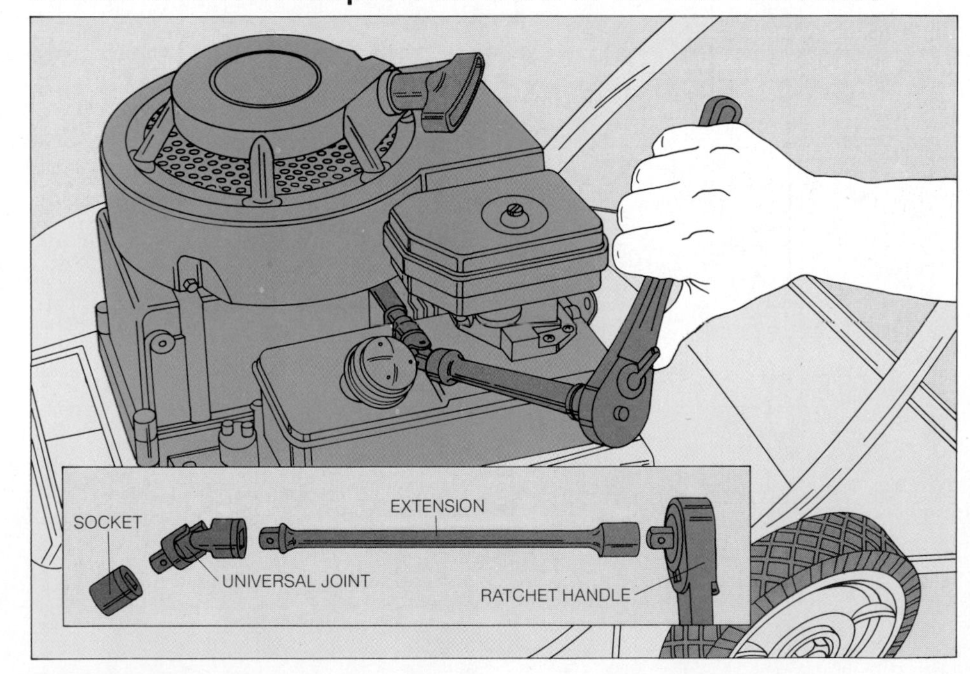

Extending the handle. To reach a bolt or a nut that is partially obstructed or recessed on the engine, add adapters to the ratchet handle. First, snap on an extension long enough to reach the fastener. Then add a universal joint to the extension and a socket to the universal joint, to complete the four-part tool *(inset)*. If the added play in the joints makes the wrench wobble as you turn it, steady it with your free hand.

A Wrench That Records the Force of a Turn

Using a beam torque wrench. To apply a precise amount of torque to a fastener, attach a socket to the torque wrench and fit the socket over the nut or bolt to be tightened. Begin to pull slowly and steadily at a right angle to the handle, taking care not to touch the flexible torque beam or the pointer arm. The beam will deflect, or bend, as you pull on the handle; the scale, attached to the beam, will move underneath the pointer. Continue pulling until the pointer reaches the desired reading on the scale.

The torque-wrench scale shown here *(inset)* gives torque in both inch-pounds and foot-pounds. If your scale shows only foot-pounds, you can translate the reading into inch-pounds by multiplying each number by 12. To convert inch-pounds to foot-pounds, divide by 12.

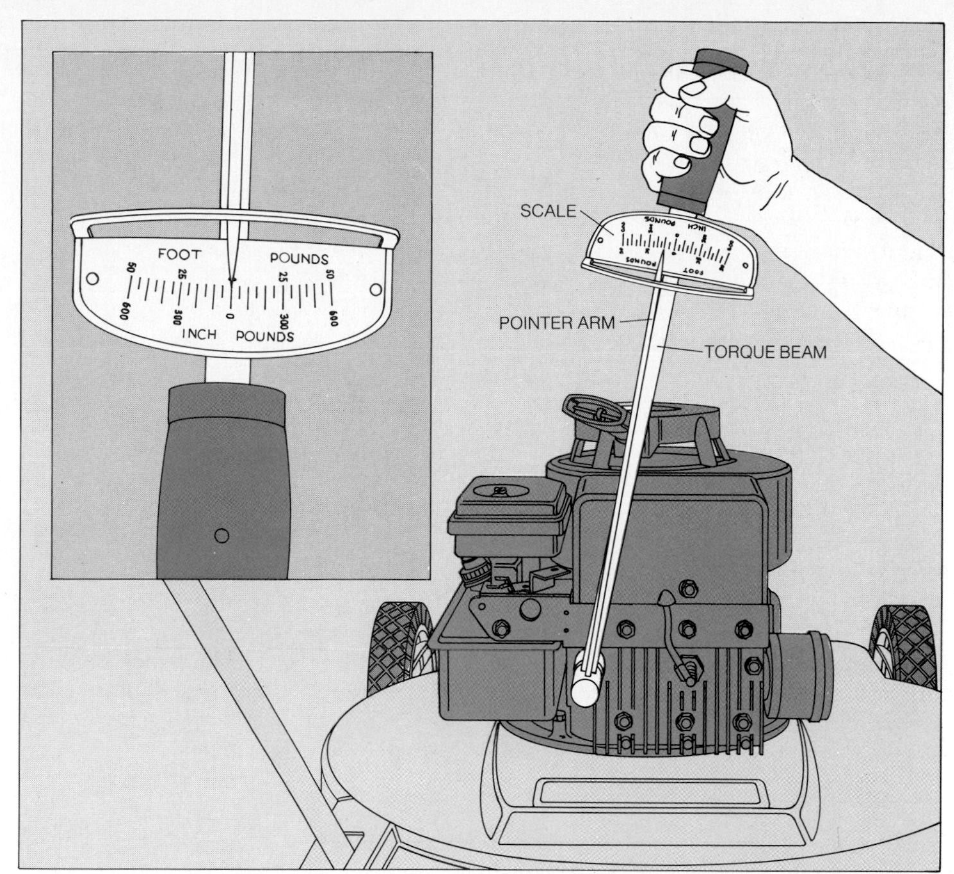

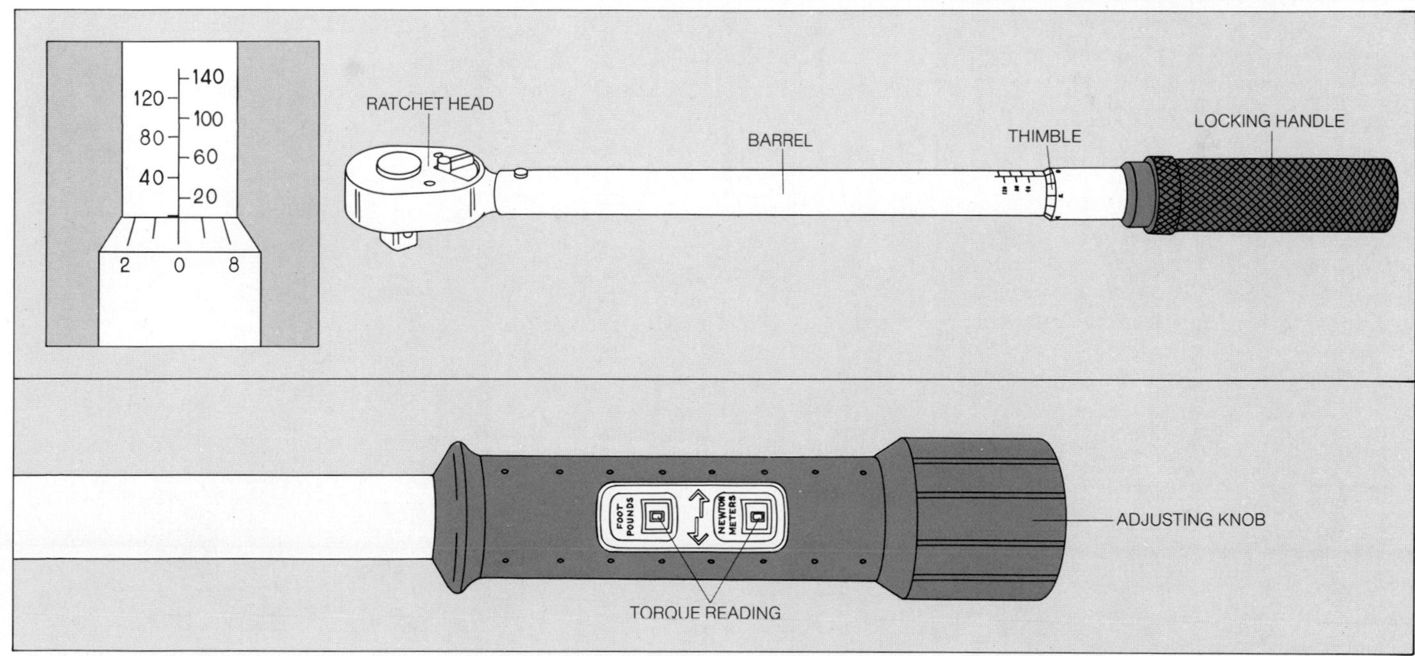

A wrench that sounds a signal. This type of torque wrench is designed so that it can be preset to give an audible clicking signal when the desired torque reading is reached; on some models, the sound is followed by a brief release of tension. To set the torque on the model shown at top, above, you must first pull out the end of the handle to unlock the wrench, and then rotate the handle clockwise so that the thimble edge moves over the scale on the barrel *(inset)*. If you want to dial the torque to 66 foot-pounds, for example, first line up the thimble edge with the 60 mark, then turn it partway around so that the minor graduation mark on the thimble itself is at 6. Finally, lock the setting by sliding the handle back up the shaft.

On a second type of click torque wrench *(above, bottom)*, the torque readings are displayed on the handle itself, here in both foot-pounds and newton-meters. To set the torque, pull out the adjusting knob at the base of the handle and turn the knob until the desired torque reading appears in the window. Lock the setting by pushing the knob back into the handle.

Tools That Measure in Thousandths of an Inch

Vital parts of small engines are made to very exacting specifications, calculated in thousandths of an inch. They have to be. Just four thousandths of an inch of wear in a crankshaft, cylinder or piston ring, for example, can cause the engine to run inefficiently. For measuring such precise tolerances—a step often necessary in engine repair—you need a few finely calibrated measuring instruments.

Chief among them, for adjusting the narrow gaps between breaker points and between valves and valve lifters, should be a feeler gauge, 15 to 30 thin, flat blades—of tempered steel or of plastic—screwed together in a case *(below, left)*. Plastic gauges are less expensive than metal and, although they do not stand up as well to parts-cleaning solvents such as kerosene, they have the added advantage of being nonmagnetic—a convenience when you are working near the strong magnets of a flywheel magneto.

Feeler gauges can be used to measure spark-plug gaps, but there is a danger of angling the blades incorrectly, especially if the electrodes are worn and rounded; this could cause a wrong reading. It is better to use a spark-plug gap gauge. A set of L-shaped wires of different diameters attached to a screw, this gauge gives the same reading at any angle.

Spark-plug gauges and feeler gauges are calibrated in both English and metric dimensions. They require only minimal care: frequent cleaning with a soft cloth to remove dust and corrosion.

You may eventually need to calculate the thickness and wear of critical engine parts such as valve stems, crankshafts and pistons. Such measurements are made with an outside micrometer, an adjustable device resembling a C clamp but with a scale accurate to .001 inch.

The micrometer is fitted around the part, and the reading is taken from the scale *(inset, opposite)*. For small engines the most useful micrometer is a 1-to-2-inch, the size used for measuring most crankshafts and connecting rods; for jobs such as determining valve-stem and cylinder diameters, you might consider adding a 0-to-1-inch and a 2-to-3-inch instrument. Metric micrometers measure distances in hundredths of a millimeter.

To measure an inside diameter, such as that of an engine cylinder, a T-shaped telescoping gauge *(opposite, bottom)* may be used with the outside micrometer. Telescoping gauges are sized in 1-inch increments, but they lack the micrometer scale and are thus simpler and less expensive. Using one together with an outside micrometer is cheaper than buying an inside micrometer.

Because of their precision and cost, telescoping gauges and micrometers demand special care. Using them in extreme heat or cold will cause expansion or contraction and give an inaccurate reading; and dropping them can spell the end of their accuracy.

Gauges for Tiny Gaps

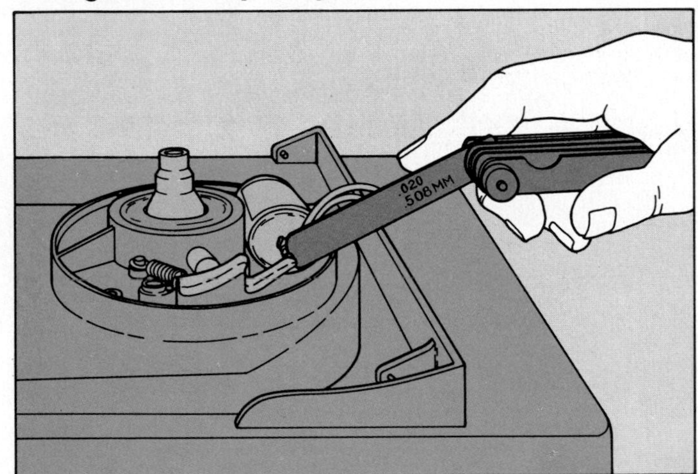

Using a feeler gauge. To determine the size of a gap, insert a blade of the feeler gauge into the opening; to get the feel of a loose fit, use a size thinner than the recommended one. Then slip progressively thicker blades into the gap until one fits snugly. The blade should not slip easily, but neither should it stick. If it does stick, the blade is too thick. The size marked on the blade that fits best is the size of the gap.

If your gauge set does not have a blade the exact thickness of the gap, stack one blade on top of another. For instance, add blades marked .015 and .012 for a total thickness of .027 inch.

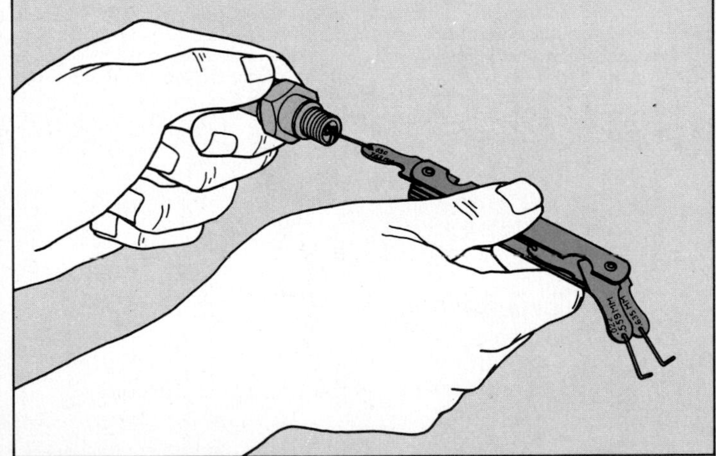

Using a spark-plug gauge. To ensure accuracy, measure the gap between the electrodes with a wire spark-plug gauge instead of a feeler gauge. Insert the gauge wire of the recommended diameter into the gap. If the wire either catches or slips out easily, the gap is incorrect and must be adjusted *(page 41)*.

Measuring Inside and Outside Diameters

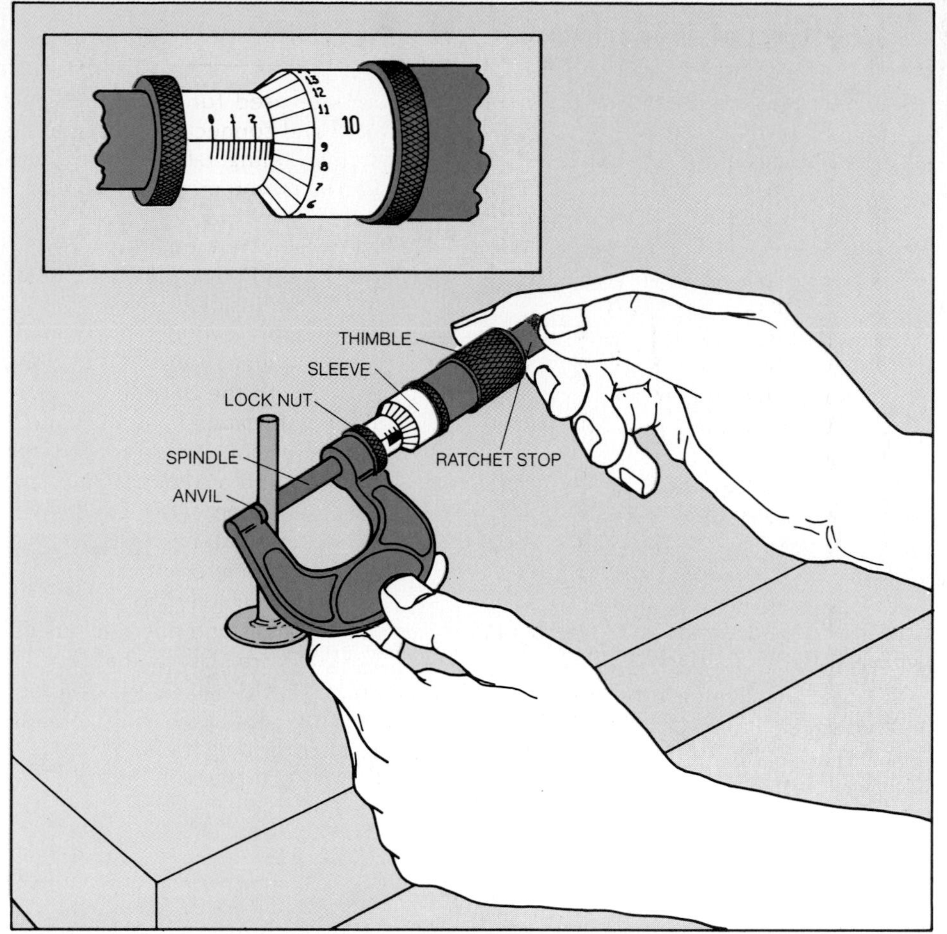

Measuring an outside diameter. Place the anvil of the micrometer against the part to be measured, and turn the thimble until the spindle almost touches the part. Then turn the ratchet stop until it clicks—this is the signal to stop tightening the spindle.

To read the micrometer scale *(inset)*, note where the edge of the thimble stops along the scales, and add the two sleeve readings to the thimble reading: The sleeve is numbered in major gradations of .100 inch, subdivided by marks into increments of .025 inch; the thimble's vertical scale breaks the measurement into even smaller units of .001 inch. In the example of the 0-to-1-inch micrometer shown in the inset, the edge of the thimble stopped at the third .025 mark—or .075—past 2 on the sleeve scale, so the sleeve reading is .2 plus .075, or .275. The vertical thimble scale is at 9, which indicates .009. Adding the .009 from the thimble scale to the .275 from the sleeve scales gives a total measurement of .284 inch. A larger micrometer, such as a 1-to-2-inch model, would be read in the same way, but the total measurement would be 1.284 inches, since the scale starts at 1 inch.

If the part is too far inside the engine for you to see the scales easily, tighten the lock nut to secure the spindle and the readings, and then carefully remove the micrometer.

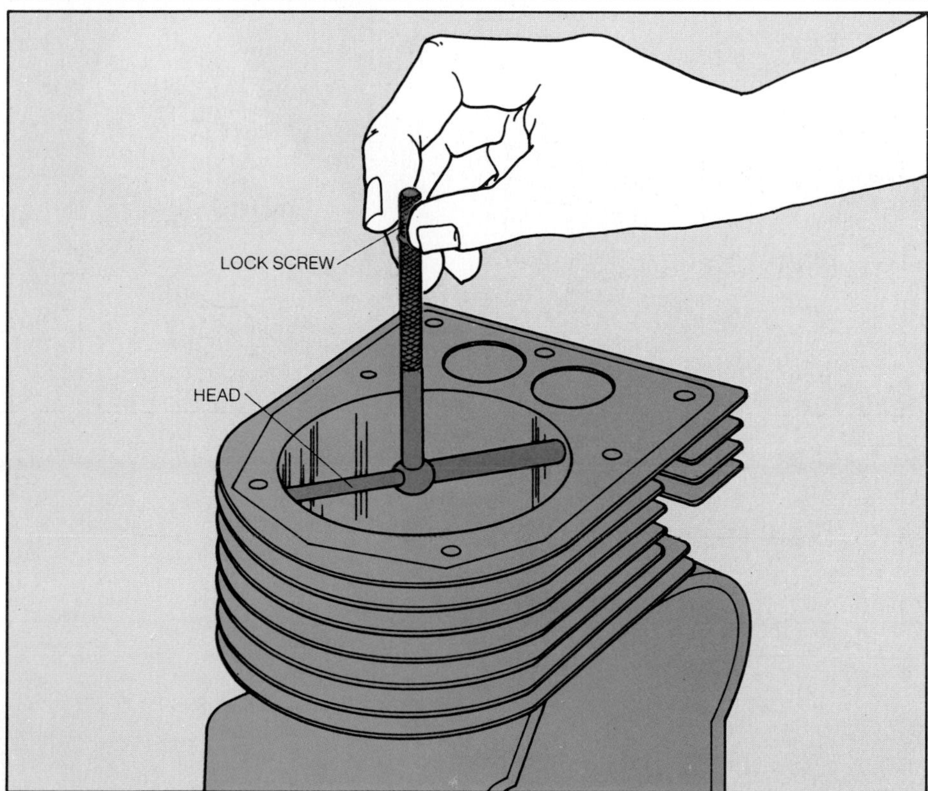

Measuring an inside diameter. Squeeze the spring-loaded head of a telescoping gauge until the head is visibly shorter than the inside diameter of the part you will measure; then turn the lock screw to secure the head. Insert the head into the part and release it by loosening the lock screw—the head will pop out and touch the inner walls of the part, spanning the inside diameter. Tighten the lock screw again, then test for the true diameter of the part by rotating the head: If it sticks or reveals gaps, refit the gauge until you have found the exact inside diameter. Then pull the gauge out and measure the length of the head with an outside micrometer; this distance is the inside diameter of the part.

Holding an Engine Together

The all-important task of holding a gasoline engine together falls to an array of specialized machine fasteners—screws, bolts, nuts, washers and devices called keys. Though similar in appearance to the familiar wood fasteners used by carpenters, machine fasteners are different in a number of ways.

Because machine fasteners must withstand the poundings and vibrations of an engine at work, the steel used to make them is hardened and strengthened by a heat process called tempering. In addition, the threads on a machine fastener are more finely cut and more closely spaced than those on a wood fastener. Extra threads per inch provide more surface area; more surface area ensures a tighter grip. Washers and nuts help by filling small gaps and by locking fasteners snugly in place.

Yet despite their strength and holding power, machine fasteners still come loose, still become worn and stripped. When this happens they should be rethreaded or, in extreme cases, replaced. The tools used to remove and replace machine fasteners are described in the home mechanic's tool kit on pages 8-9. To ensure an exact match for an old fastener, take it with you when you buy new parts. The first place to go is to the dealer from whom you bought the engine. If the fastener you need is not in stock, try a large hardware store or an automotive-supply store.

A gallery of fasteners. Machine screws are the most common engine fasteners. They come with a variety of head shapes—oval, hexagonal, pan and fillister—which are slotted to accommodate either flat-tipped or Phillips screwdrivers. Machine bolts are larger than screws and are used for fastening heavier engine parts. They have square or hexagonal, unnotched heads designed for wrenches rather than screwdrivers. Setscrews, which join movable parts, are headless; their tops have openings for hex wrenches. Thumbscrews have flat or winged heads, enabling them to be turned by hand. Narrow sheet-metal screws with pointed tips bore through lightweight, unthreaded metal parts. Both screws and bolts are used with a variety of flat and locking washers; both can be used with either square or hexagonal nuts.

A key is an unthreaded, semicircular fastener of mild steel. It may fit, for example, into a slot in the crankshaft and a groove in the flywheel *(inset)*, holding the two together. If the flywheel jams, this relatively soft metal fastener will break, letting the crankshaft turn independently so that the two parts will not damage each other.

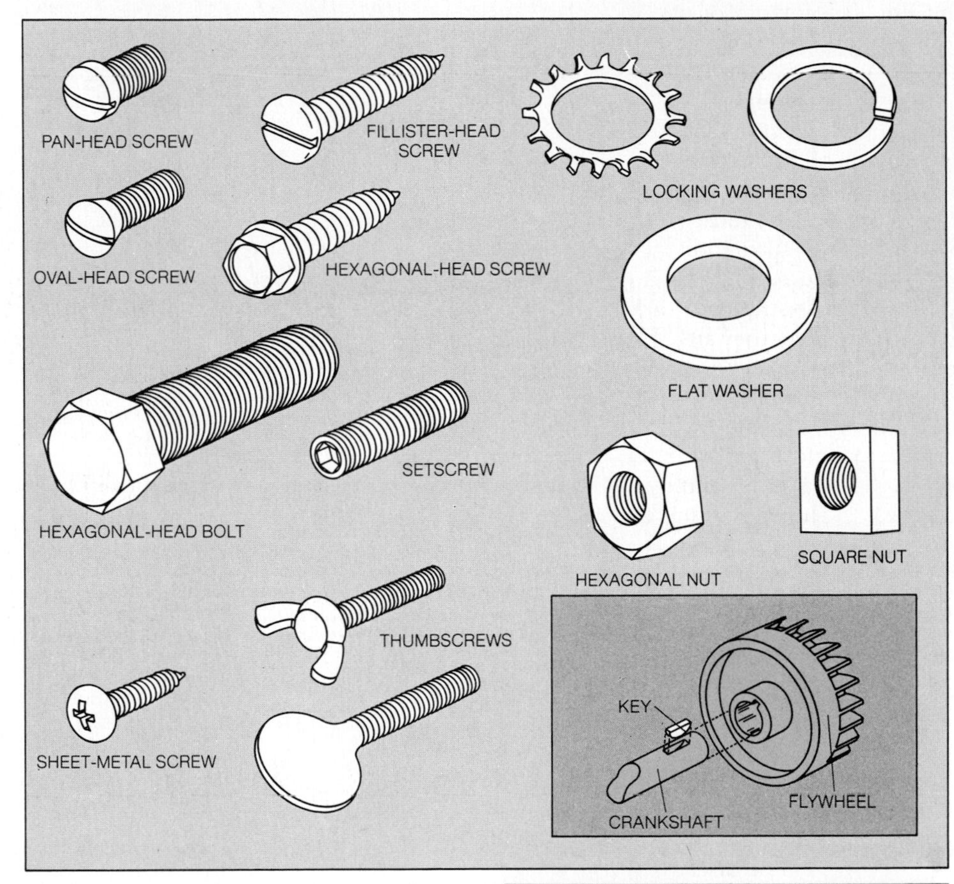

A fastener's vital statistics. Machine screws and bolts are identified according to three dimensions:the diameter of the shaft at the widest part of the threads; the number of threads per inch; and the length of the shaft from the tip to just below the head, except for oval- and flat-head screws, whose length includes the head. If an engine manual calls for a bolt, followed by the figures 1/4-20-1, it recommends a bolt with a shaft that is ¼ inch thick, has 20 threads per inch and is 1 inch long. Nuts and washers for machine fasteners are identified according to the diameters of their inside openings.

Fastener sizes are standardized throughout the United States and Canada, but the fasteners shown here are not interchangeable with those classified according to the metric system.

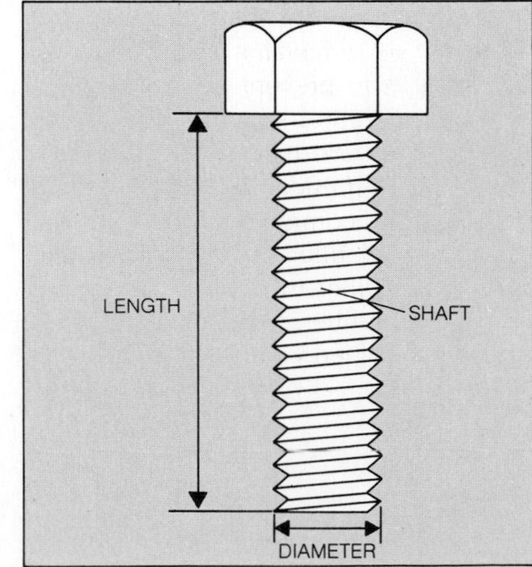

Keeping Fluids in Their Places

Gasoline, oil and water—the liquids inside an engine that fuel, lubricate and cool it—all have their prescribed chambers and channels, and for the engine to run smoothly these vital fluids must remain where they are supposed to be. The job of keeping fluids from leaking out, and keeping dirt or other impurities from seeping in, belongs to an assortment of engine sealers known as gaskets, oil seals and O-rings.

The majority of the sealers in an engine are gaskets—thin pieces of material resilient enough to fill any tiny gaps between mated metal parts; most are made of fiber or, where surrounding parts reach high temperatures, of metal and asbestos. All gaskets should be replaced, not reused, whenever they are dislodged during a maintenance or repair job.

Gaskets are sometimes used in conjunction with sealants—which act as adhesives to hold a gasket in place while you fit parts together and then as fillers, oozing into and smoothing irregularities on metal surfaces to ensure a leakproof joint. The best sealants are ones that are nonhardening and can be applied with a brush or some other small applicator; aerosol sealants, while convenient, are difficult to aim accurately. Spread any sealant as thinly as possible over the mating surface of the part, avoiding any holes for screws or bolts.

Oil seals are made of circular steel springs encased in rubber. Like gaskets, these too should be discarded once they are dislodged; even if the spring has not broken during removal, it will have loosened enough to prevent a tight seal.

O-rings, doughnut-shaped seals made of rubber that are recessed in grooves, may be reused if undamaged. These sturdy sealers are commonly found at the base of the oil filler cap, holding it securely to the filler tube. Because of the resilience of rubber, it is unnecessary to use a sealant when you install an O-ring.

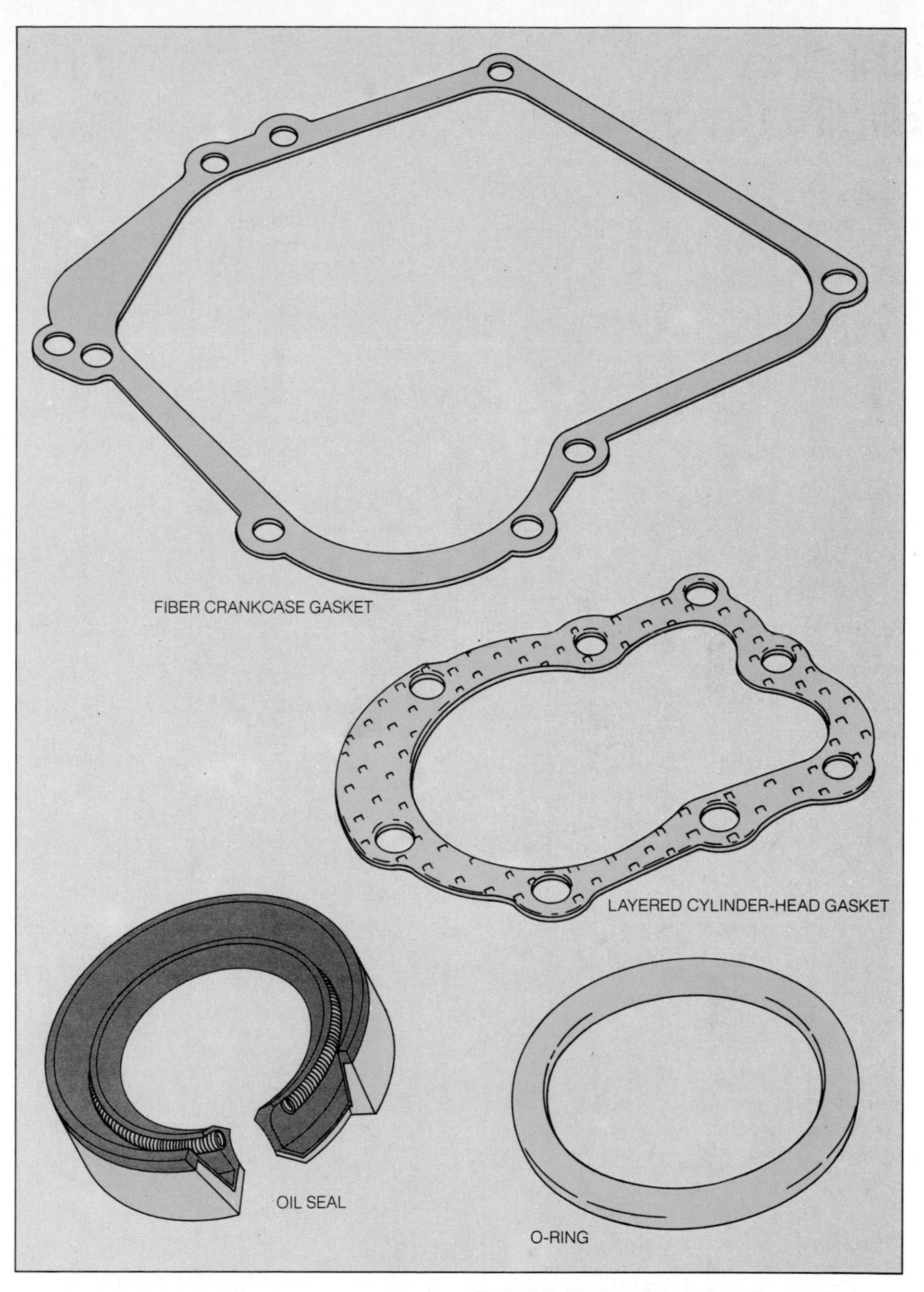

Sealers in all shapes and sizes. Sandwiched tightly between stationary parts, gaskets fill in scratches, dents and warps, sealing the space between metal surfaces. Fiber gaskets are used where the mating surfaces are uniform, such as the crankcase cover; plastic spacer blocks line some fiber gaskets. In spots where pressure and temperature are high, such as the cylinder head, gaskets of layered metal and asbestos are used. Oil seals are used around moving parts—both ends of the crankshaft, for example. A steel spring grips the end of the shaft, and a rubber lip surrounding the spring completes the seal. Rubber O-rings fit snugly into circular grooves between engine parts, as between the carburetor and the intake manifold.

Freeing a Stuck or Broken Fastener

Rusted bolts, worn threads and mangled screw slots can bring the simplest repair job to a halt. A fastener may have been damaged by repeated attempts to remove it, or it may simply break. Mechanics overcome these frustrating obstacles with a variety of standard tricks: using penetrating oil, heat and, when all else fails, mechanical extractors.

When a fastener stubbornly refuses to turn, despite a reasonable effort with a wrench or a screwdriver, penetrating oil is the first thing to try. Thinner than ordinary oil, penetrating oil is better able to seep into spaces around fastener threads. A few drops placed under the fastener head and on any accessible threads usually do their work within five minutes. Rap the fastener repeatedly with a hammer to help the oil work down through the rust and along the threads.

If oil does not help, heat may—provided the fastener is not located near the carburetor, gas tank or fuel line, where gasoline vapor might ignite. The heat causes the metal surrounding the fastener to expand outward, thus freeing the fastener. Using a small propane torch, play the flame on the area around the fastener—but not on the fastener itself—until the heated spot warms to about 250° F. Judge the temperature by sprinkling a little water on the spot: if it sizzles, the metal is hot enough.

Sometimes fasteners get stuck so tightly that common wrenches and screwdrivers cannot budge them without the aid of a tool called an impact driver *(below)*. Made of steel, the driver converts a hammer blow into a rotary motion to loosen a screw, bolt or nut. It comes with either a ⅜-inch or ½-inch square drive to hold special impact sockets made to stand up to this hard use. A screwdriver adapter enables the tool to loosen screws. To avoid eye injury, wear goggles when using the impact driver.

From time to time, your best efforts to remove a stuck fastener may fail: The threads, head, slot or shaft of a fastener can be damaged beyond the point of usefulness. These parts of a fastener must all be dealt with differently. If the threads on a bolt have been stripped but the threads inside the hole are intact, you can simply replace the bolt with a new one. If a good bolt that is easily extracted will not tighten, the hole threads are damaged. The hole can be enlarged and rethreaded. A kit containing replacement threads even makes it possible to reuse the same bolt *(opposite, bottom)*.

In the case of damaged fastener heads, replacement of the fastener may be necessary. If a screw slot is mutilated, it may be possible to fashion a temporary slot so that the screw can be worked out *(opposite, middle)*. The next-best alternative is to obliterate the head of the screw with a drill and extricate the remnants of the shaft with pliers. A simpler method is employed on a bolt that has sheared off. After the bolt shaft is drilled out, a tool called a screw extractor *(opposite, top)* is inserted to withdraw the bolt fragment.

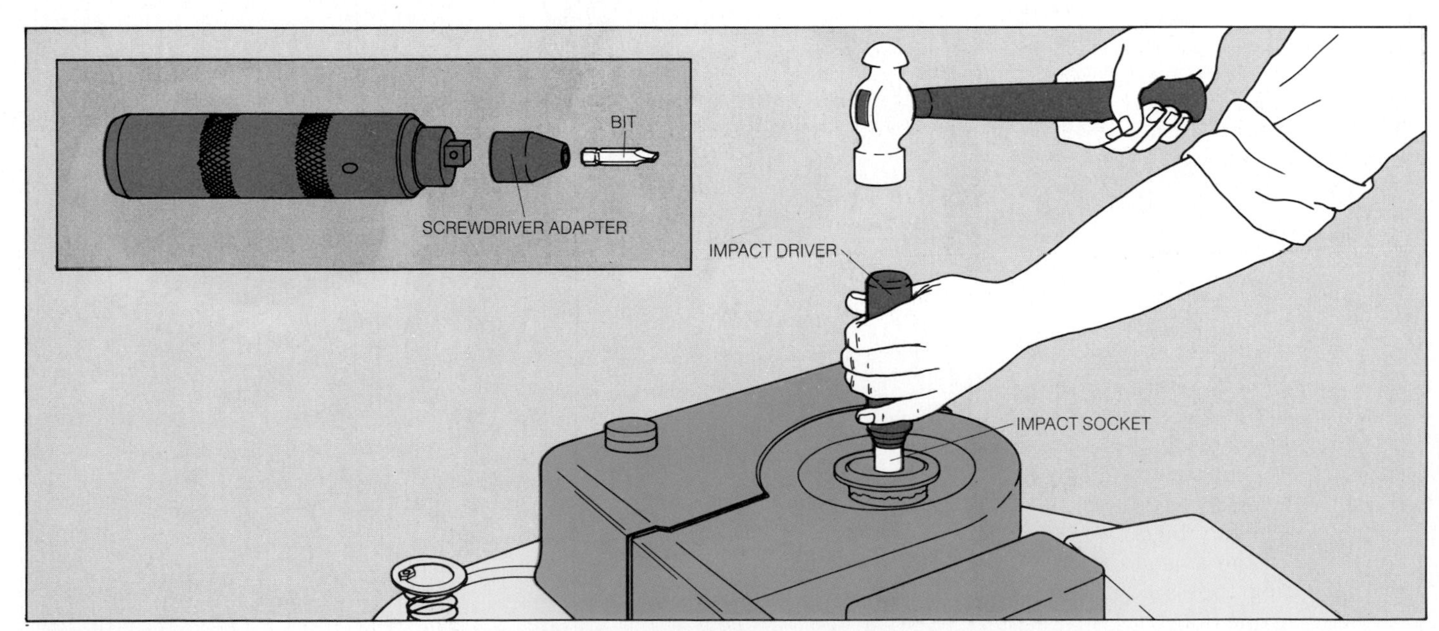

Using an impact driver. For a bolt, snap onto the impact driver an impact socket that fits the bolthead, and rotate the driver handle counterclockwise. Wearing goggles, place the socket on the bolthead and strike the end of the impact driver smartly with a 16-ounce ball-peen hammer. After each blow, rotate the handle counterclockwise to make sure that the blow has not reversed the tool's driving direction. Continue until the fastener is loose.

To loosen a Phillips or slotted screw, attach a screwdriver adapter *(inset)* to the impact driver, and insert the appropriate heavy-duty screwdriver bit. Set the driver handle and strike the driver as you would for a bolt.

Removing a Ruined Screw

Turning out a broken bolt. To remove a broken bolt with a screw extractor, first drill a hole into the bolt shaft; use a center punch to dimple the center of the broken bolt's shank to keep the drill from wandering. Measure the diameter of the shank with a ruler or a micrometer, and select a drill bit of the size recommended by the extractor manufacturer. Drill ½ inch into the shank (1 inch if the bolt is wider than ⅜ inch), oiling the bit every 10 to 15 seconds to keep the bit from overheating. When the hole is complete, lubricate the bolt threads with a few drops of penetrating oil. After five minutes, put the extractor into the hole and turn it counterclockwise with a tap wrench, or an open-end or adjustable wrench, twisting out the broken bolt.

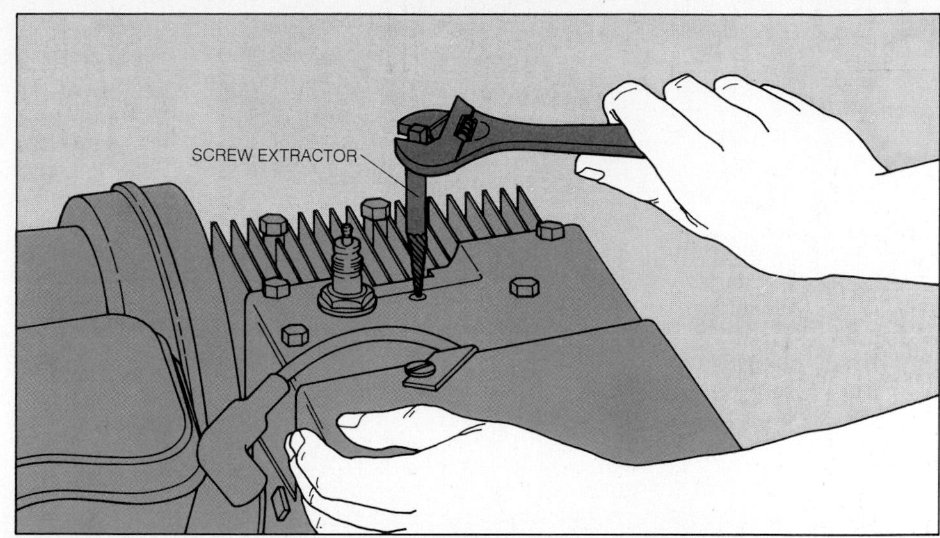

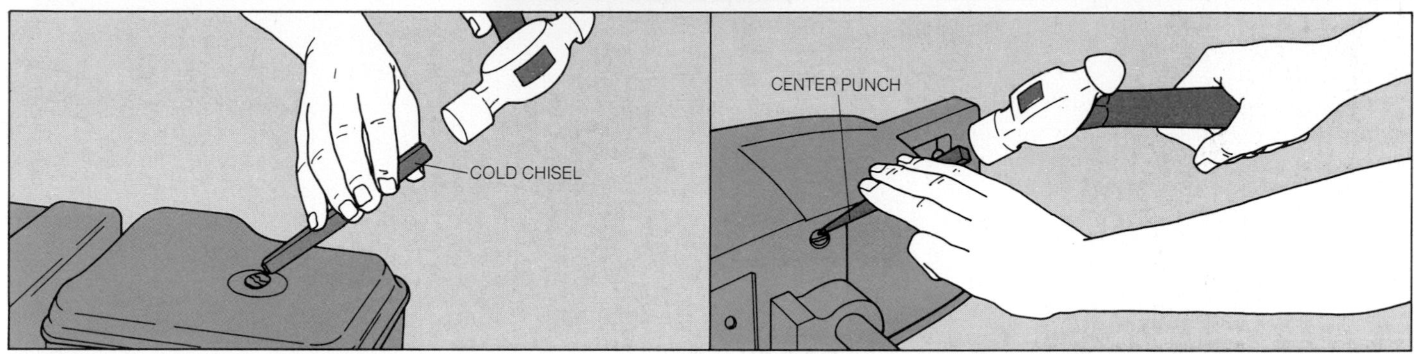

Creating a new slot. To remove a screw that has a head protruding above the surface *(above, left)*, notch the side of the screwhead with a small cold chisel and a ball-peen hammer. Keeping the chisel in the notch, turn the screw counterclockwise by continuing to tap the chisel with the hammer. The screwhead will rise above the surface; when it is about ⅜ inch high, grip the head firmly with locking-grip pliers and unscrew it the rest of the way.

For a screw that is flush with or below the surface *(above, right)*, drive a center punch into a thick part of the head, avoiding the slot. Hammer the punch straight into the head, then angle the punch to the right and tap the screw counterclockwise to remove it. If these two methods fail, drill off the head of the screw with a drill bit slightly smaller than the screwhead. Remove whatever the screw was holding and extract the exposed shank with locking-grip pliers.

Rethreading a Stripped Bolthole

A bolthole needs rethreading when a bolt in good condition will not tighten or threads are obviously missing from the hole. To cut new threads into a hole, it is necessary to drill out the hole to make it slightly bigger than before and rethread the enlarged hole with a thread-cutting tool called a tap. This method has one drawback: you need a new, slightly thicker bolt for the new hole size. There is, however, a method that allows you to reuse the original bolt; it calls for a special kit containing replacement threads.

To use the kit, you begin by measuring the size of the fastener and boring a slightly larger hole to the original depth. This hole is then threaded—or tapped—with a tapping tool.

Next, a thread insert is turned into the hole, using a metal rod called a mandrel, which has a catch at the end to rotate the insert's end coil, or tang. A plastic sleeve, or driver, is slipped over the mandrel to steady it as it is turned by a wrench. The top of the insert is driven to just below the top of the hole, and the mandrel is removed. A center punch is used to break the tang, so the fastener can enter the hole.

Thread-insert kits are sold at auto-supply stores. However, because a different kit is needed for each size fastener, it may be more economical to have a bolthole rethreaded at a machine shop.

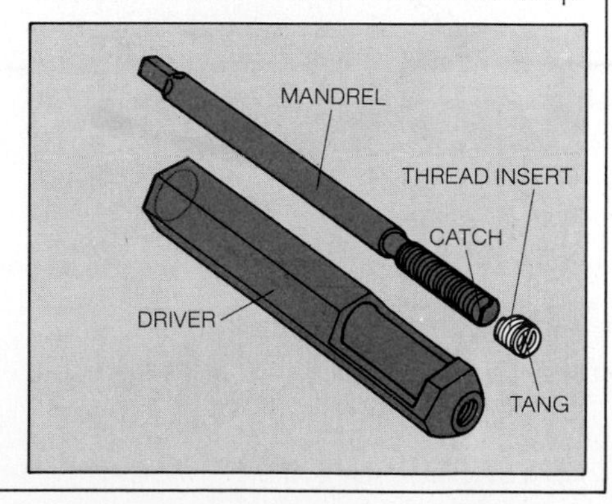

.020
.508 mm

2 The Art of Everyday Maintenance

Setting the gap. The leaf of a feeler gauge, selected from leaves of varying thickness, is used in setting the distance between two electrical contacts—in this case, at exactly .020 inch. The contacts, called breaker points, control surges of voltage to the engine's spark plug, and the distance between them is critical for the correct timing of the spark. Such precise adjustments in the ignition system are perhaps the most important part of an engine tune-up.

The quality of an engine's work and the length of its working life depend not only on its design and materials, but on the care it receives. Routine maintenance and regular tune-ups will keep it running smoothly for years, and both are well within the power of a home mechanic. Maintenance involves little more than common-sense attention and service—periodic inspections, cleaning and lubricating. A tune-up, which calls for adjusting the fuel and ignition systems and regulating the timing, is more complex but scarcely more difficult. Both are based upon proven techniques that have been refined over more than a century of small-engine development.

Home mechanics work with two categories of engines, two-stroke and four-stroke *(pages 20-25)*, whose designs date back to the 19th Century. At the Philadelphia Centennial Exposition of 1876, George Brayton, a Boston engineer, exhibited the first liquid-fuel engine—a large, languid two-stroke machine that ran on kerosene. In the same year, a German engineer named Nicholas Otto designed and built a gasoline engine with a four-stroke cycle of operation. A wide variety of customers made the Otto engine an immediate success. Machinists used it to run lathes; brewers put it to crushing hops.

Today's small engines are faster, lighter and more efficient than their ancestors, but they operate on the same principles and have much the same mechanical parts. To the uninitiated, the engines of a rotary lawn mower, a garden tiller and a small electric generator may look entirely different; actually they all work the same way and may even have rolled off the same production line.

A small-engine manufacturer often produces one basic model of an engine—a 3½-horsepower, four-stroke model, for example, making only exterior modifications to fit it to a particular piece of machinery. The most obvious differences will be in the location and orientation of various external parts. For example, the fuel tank and carburetor on a lawn-mower engine, which is mounted vertically, will be located differently from those of a rotary-tiller engine, which has a horizontal alignment. But aside from the arrangement, and perhaps the shape, of their components, the only major distinction between the two will be in the ends of their crankshafts, which are adapted to the business ends of these machines. The crankshaft on the tiller may be fashioned with a channel for the insertion of a key; on the mower, the engine shaft may be drilled with a threaded bore for the bolt that holds the blade.

In caring for small engines, though, the differences are far less significant than the similarities. Once you are familiar with the systems that make the two basic varieties work, you will be able to find your way around any small engine with ease and tune it with assurance.

Getting Useful Power from a Four-part Cycle

Small gasoline engines have changed little since their introduction late in the 19th Century. Now, as then, the core of the small internal-combustion engine is a metal cylinder that houses a close-fitting drum called a piston. The piston, which moves from one end of the cylinder to the other, is attached by means of a jointed steel limb to a revolving shaft called the crankshaft. In turn, the crankshaft drives the business end of the tool—such as the blade of a lawn mower or the drive sprocket of a chain saw.

In order to produce power to turn the crankshaft, an engine goes through four distinct stages. First, the fuel charge—a vaporous mixture of gasoline and air—is admitted to the cylinder. There it is compressed by the advancing piston, creating heat energy that brings it almost to the point of spontaneous combustion. Next, it is ignited by an electrical spark and expands rapidly, driving the piston away. Finally, the waste gases from the spent fuel are exhausted from the cylinder.

These four steps—intake, compression, ignition and exhaust—make up an engine's power cycle. Each step occurs during the operation of all small gasoline engines. Engines differ, however, in the number of times the piston moves from one end of the cylinder to the other in order to complete a power cycle. The piston movements are called strokes, and engines are divided into two classes: two-stroke and four-stroke, also commonly called two-cycle and four-cycle.

The four-stroke engine *(opposite)* is the simpler of the two to understand, because it uses one stroke to accomplish each of the four steps *(page 22)*. Two strokes occur during each revolution of the crankshaft, so a four-stroke engine spins the crankshaft twice before completing one power cycle.

The two-stroke engine *(pages 23-24)* completes its power cycle with only two movements of the piston—one revolution of the crankshaft. A two-stroke engine achieves its economy of motion through a design that allows two steps to occur simultaneously during each stroke: A fresh fuel mixture flows into the crankcase as the fuel already in the cylinder is compressed. Then, as the fuel is ignited and the piston is thrust back, the fresh charge is transferred into the cylinder, and fumes from the spent fuel are shunted out through an exhaust port. Because it operates without the intricate valve system essential to the four-stroke model, the two-stroke engine has fewer than half as many moving parts.

The differences in design between two- and four-stroke engines lend themselves naturally to different applications. In general, the larger, heavier four-stroke engines run cooler and more smoothly but have a lower capacity for quick acceleration or high-rpm operation. They are commonly used to power snow blowers, generators and lawn mowers—applications that require low-speed operation for extended periods.

Two-stroke engines, which are smaller and run hotter, are built for brief but demanding use. With the quick acceleration that comes from one turn of the crankshaft per power cycle, these compact power plants are used to drive chain saws, small weed-trimming garden tools and outboard motors.

The smooth performance of the power cycle in both two- and four-stroke engines depends on several intricately interrelated subsystems: fuel, carburetion, ignition, exhaust, lubrication and cooling. The fuel system transfers the gasoline from the fuel tank to the carburetor, where it is mixed with air before entering the cylinder. This vaporous mix—approximately one part fuel to 15 parts air under normal operating conditions—passes into the cylinder through an opening timed to close just before the piston begins its compressive thrust toward the spark plug. The engine's ignition system generates a high-voltage current and sends it through a wire to the spark plug to produce the spark that explodes the fuel mixture.

Without lubrication and cooling systems, this sequence could not take place: The engine would grind itself to a halt or overheat. Oil prolongs the life of an engine's moving parts by reducing friction between them; it also helps to cool the engine by absorbing heat from the parts it lubricates. In four-stroke engines, oil is circulated by a pump *(page 21)* or by a simple splash system that uses a dipper attached to the rotating crankshaft. The dipper ladles oil from a reservoir in the crankcase and then flicks it onto the engine's moving parts.

Two-stroke engines are lubricated by oil that is mixed with the gasoline before the gas is poured into the fuel tank; the oil then does its work as it courses through the engine's fuel system and cylinder, where it is eventually ignited and exhausted along with the gasoline.

Although a few small engines—outboard engines, for example—are water-cooled, most are air-cooled. Fins on the exterior of the cylinder provide a large surface area that quickly dissipates heat. Additionally, air is circulated over the engine by a fanlike flywheel *(page 24)*.

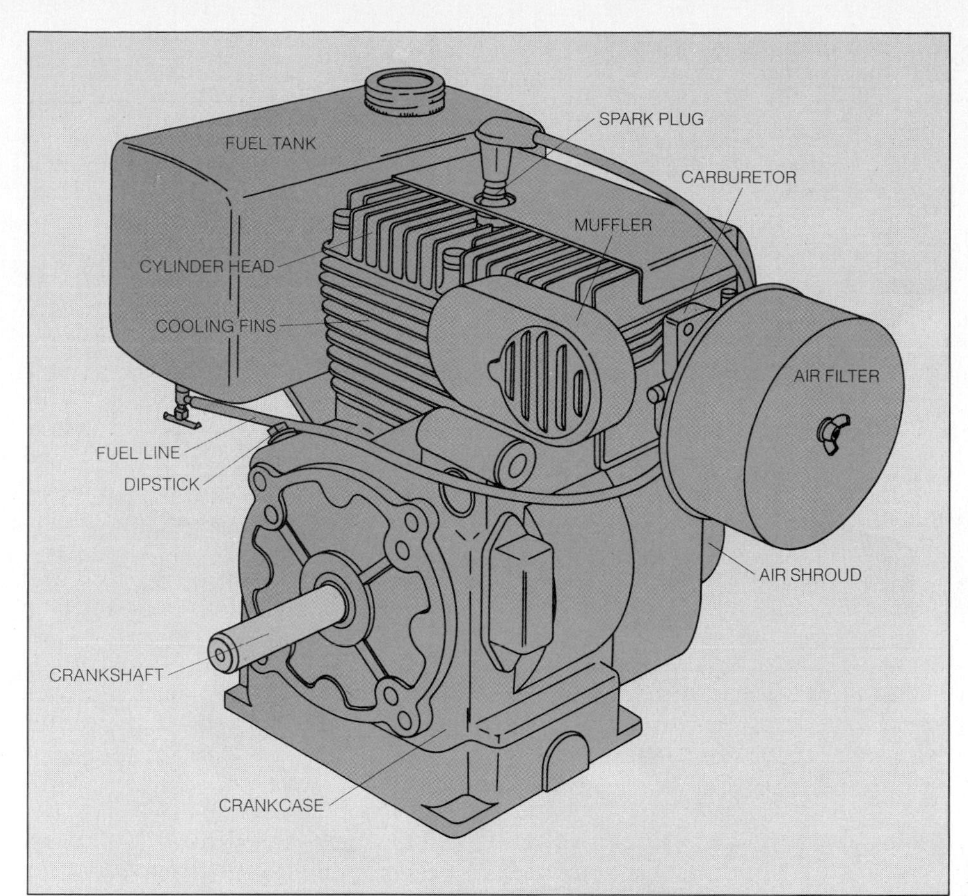

A Four-stroke Power Plant

Portrait of a four-stroke engine. The external components of a four-stroke engine are illustrated here on a model typical of those used to run garden tillers. The same parts may be sized, shaped or oriented differently on other engines, but they will all be located in similar positions relative to one another.

Gasoline from the fuel tank travels to the engine through a feeder tube called a fuel line. On its way to the cylinder, it passes through the carburetor, where it is mixed with air drawn in through the air filter. A spark plug is screwed into the cylinder head—the top of the combustion chamber in which the fuel is ignited. A muffler covers the exhaust port to damp the din of escaping waste gases. The cooling fins on the outside of the cylinder offer a large surface area that dissipates the heat generated by combustion.

A finned flywheel (covered by a metal shroud) functions as a fan to cool the engine further. The flywheel is connected to the crankshaft, which passes straight through the engine in a housing called the crankcase. (The section of crankshaft visible here is normally concealed by whatever implement the engine powers.) A dipstick, which emerges from an oil reservoir in the crankcase, has a nutlike head that secures it while the engine is operating.

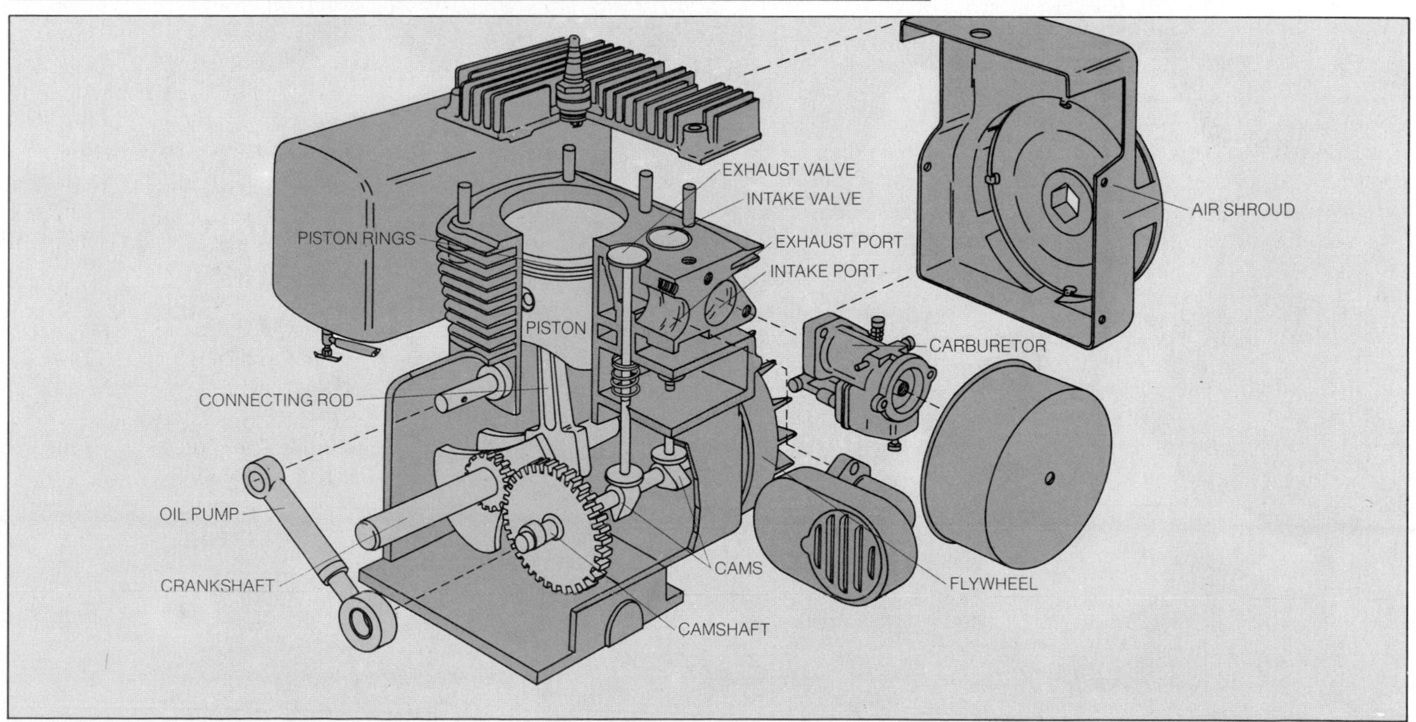

Inside a four-stroke engine. The air-fuel mixture from the carburetor passes through the intake port and intake valve into the cylinder; waste gases pass through the exhaust valve and out the exhaust port. The piston—encircled by cast-iron rings that maintain snug contact with the cylinder wall—pushes and pulls a connecting rod that rotates the crankshaft. This rotation is transferred by two gears to the camshaft, which operates the barrel-type oil pump by driving its plunger. The camshaft also raises and lowers the intake and exhaust valves by rotating the pear-shaped cams beneath the valve lifters. Momentum created by the heavy flywheel helps keep the crankshaft revolving smoothly.

The Stages of a Four-stroke Cycle

Intake stroke. The piston moves away from the spark plug, creating a partial vacuum in the cylinder that draws the air-fuel mixture in through the open intake valve, which has been lifted up by the cam beneath it.

Compression stroke. As the piston advances toward the spark plug, it compresses the air-fuel mixture, raising its temperature almost to the point of combustion. The tightly sealed valves and the cast-iron piston rings that hug the cylinder walls give the fuel mix no avenue of escape and ensure high pressure within the engine.

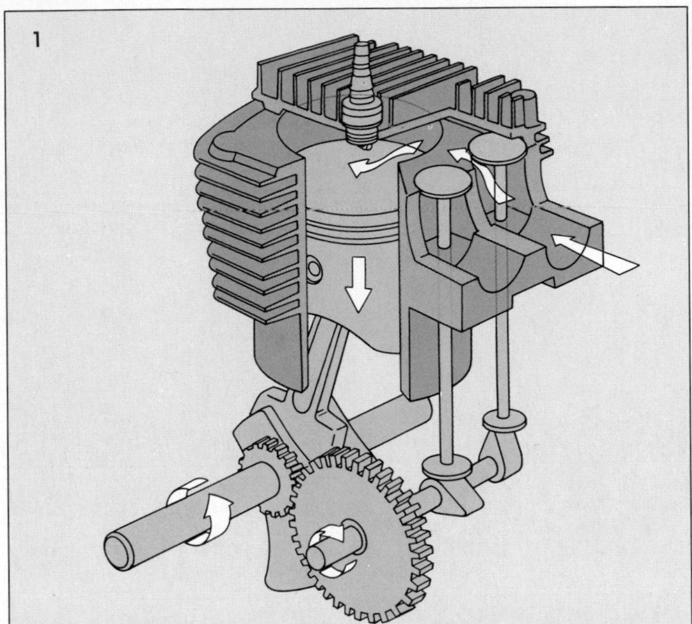

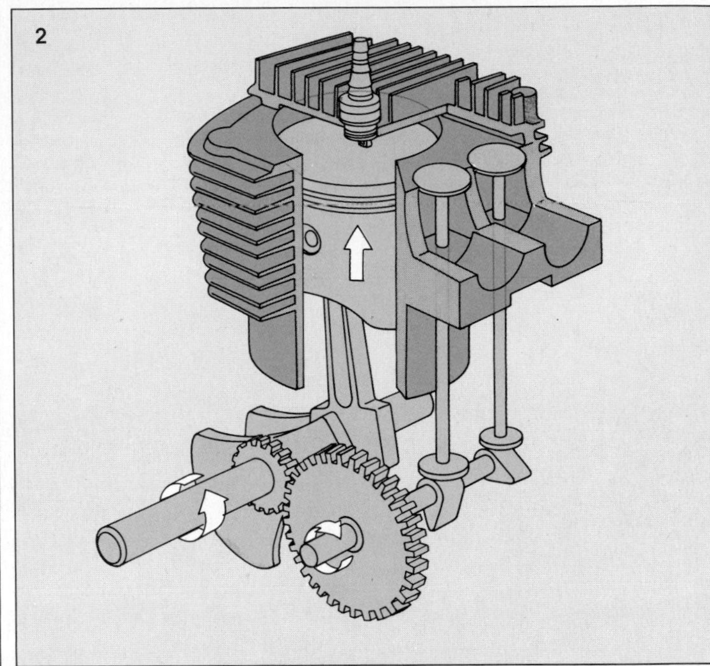

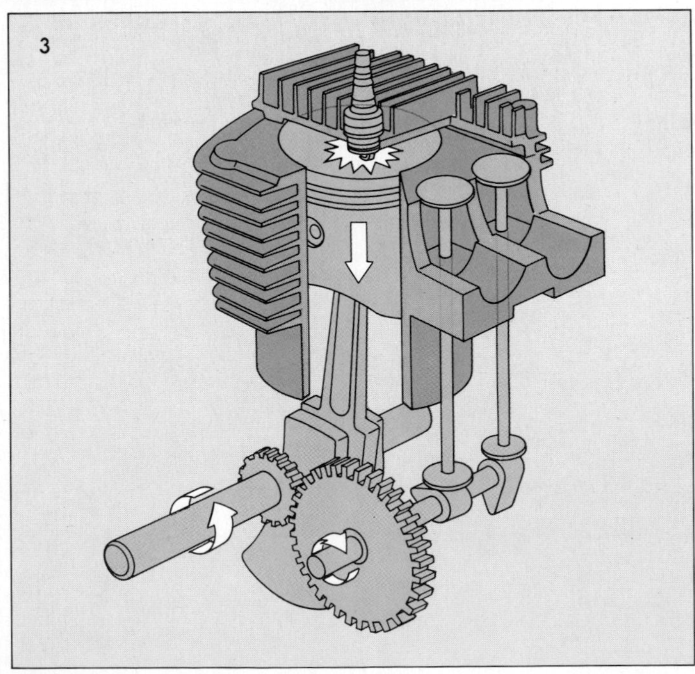

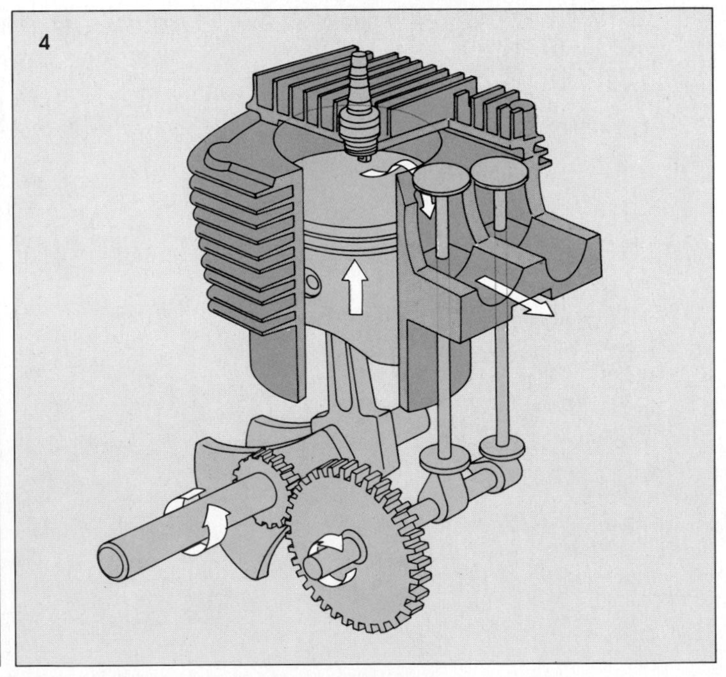

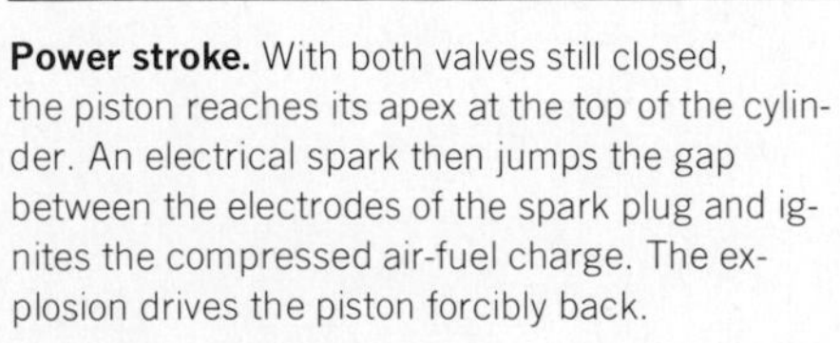
Power stroke. With both valves still closed, the piston reaches its apex at the top of the cylinder. An electrical spark then jumps the gap between the electrodes of the spark plug and ignites the compressed air-fuel charge. The explosion drives the piston forcibly back.

Exhaust stroke. As the piston rebounds, rising from the bottom of the cylinder, it pushes burned gases out through the open exhaust valve, making way for a fresh charge of air and fuel.

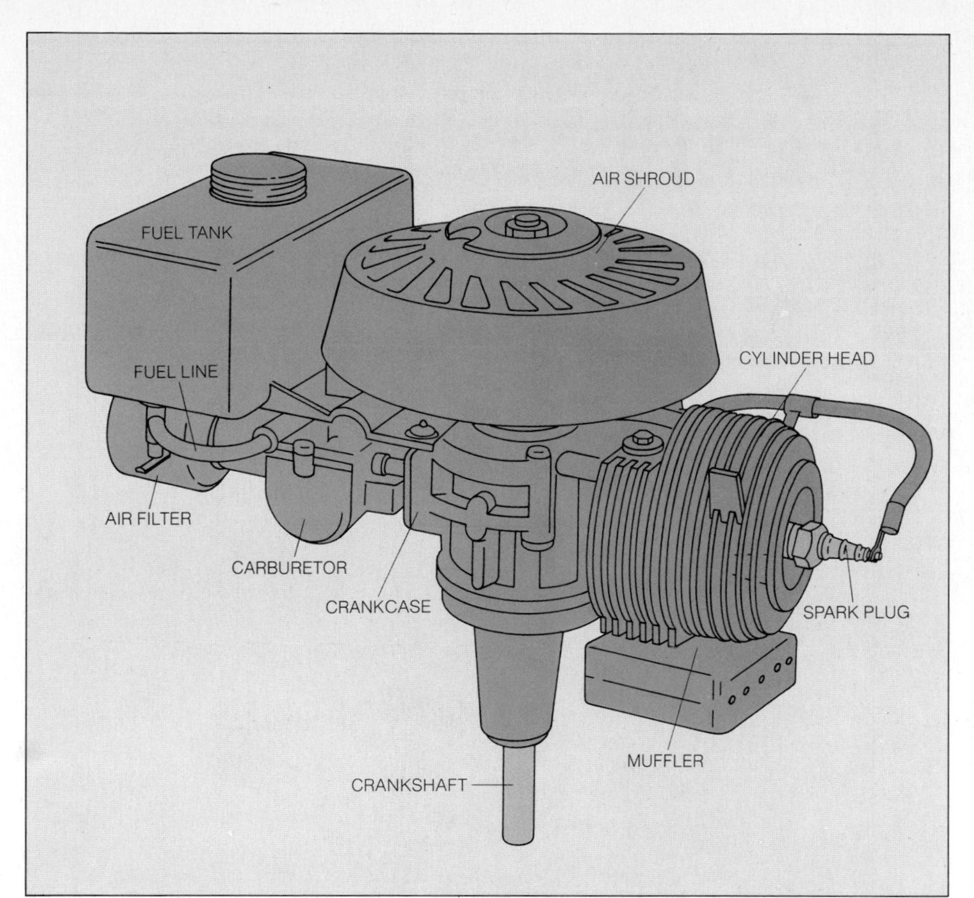

A Two-stroke Power Plant

A two-stroke engine. A two-stroke engine differs little from a four-stroke one externally. The major distinction is the position of the carburetor and air filter. Because the air-fuel mixture enters the engine through the crankcase, the intake port (and thus the carburetor and air filter) is at the base of the cylinder and crankcase, not at the cylinder head. Otherwise, the parts are identical: A flexible fuel line connects the fuel tank to the carburetor; a flywheel, covered by a metal shroud, is attached to one end of the crankshaft; a muffler covers the exhaust port; and a spark plug is screwed into the cylinder head.

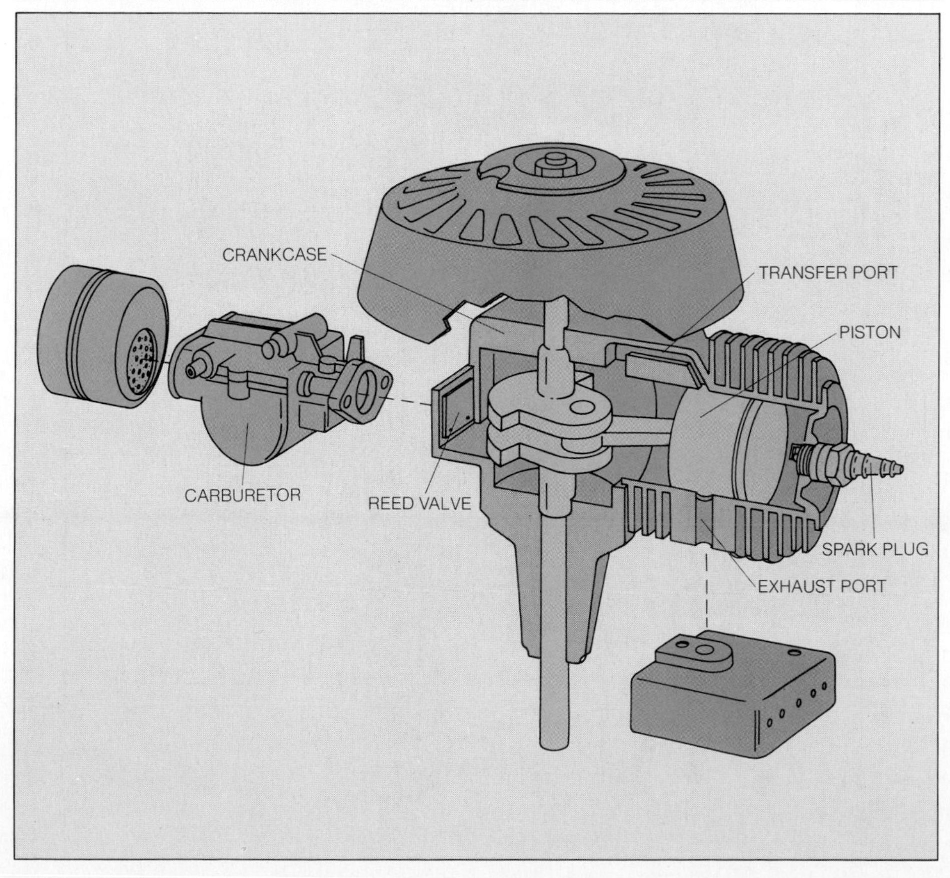

Inside a two-stroke engine. The interior of a two-stroke engine is simpler than that of a four-stroke one. There is no separate lubricating system: Oil, premixed with the gasoline in the fuel tank, lubricates the engine's moving parts as it flows through the crankcase. The air-fuel-oil mix enters the engine from the carburetor through a one-way reed valve—a flexible strip of spring steel that is drawn open as the piston moves toward the spark plug, creating a partial vacuum in the crankcase. The two-stroke engine has no camshaft and no cam-driven valves; the piston acts as a valve to open and close the transfer and exhaust ports, which admit the fuel mixture to the cylinder and expel the waste gases.

The Stages of a Two-stroke Cycle

Compression stroke. The piston moves toward the spark plug, compressing the fuel mixture in the cylinder. The advancing piston creates a partial vacuum below it, drawing a fresh fuel-and-air charge into the crankcase through the reed valve. The piston wall seals the transfer port and exhaust port shut during this stroke.

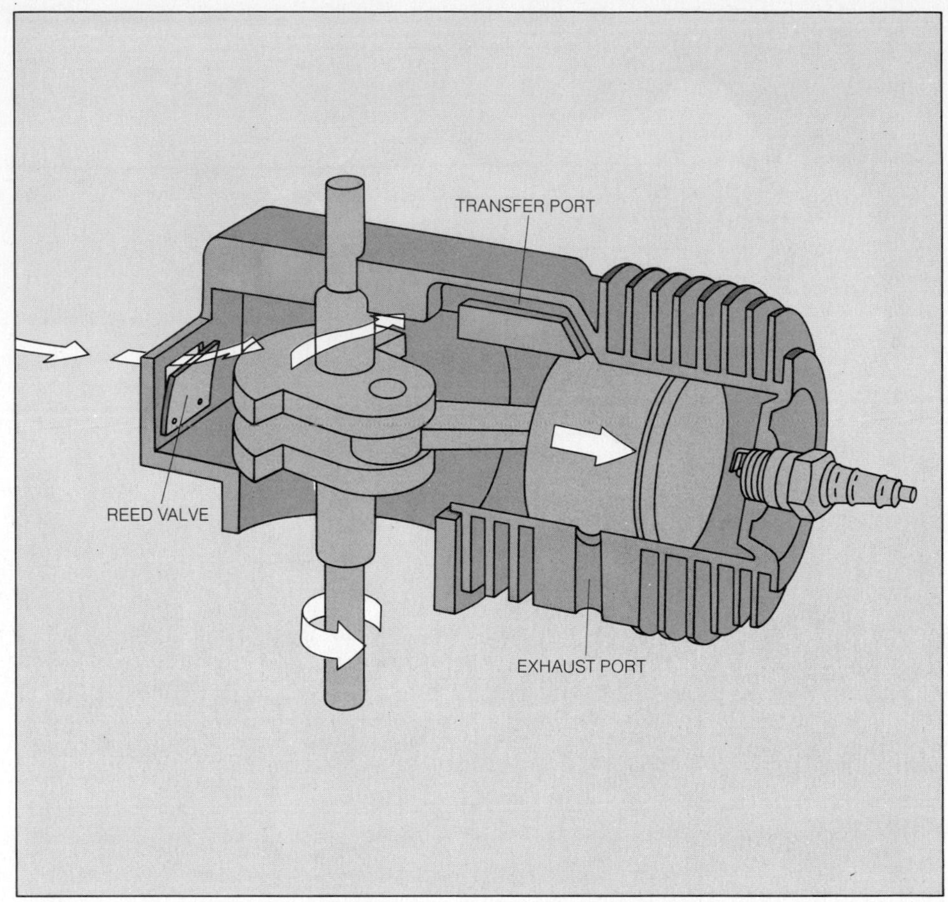

Power stroke. A spark explodes the compressed fuel mixture, causing the piston to recoil and uncover the transfer and exhaust ports. The spent fuel gases are expelled through the exhaust port, pushed out by the fresh charge entering the cylinder through the transfer port.

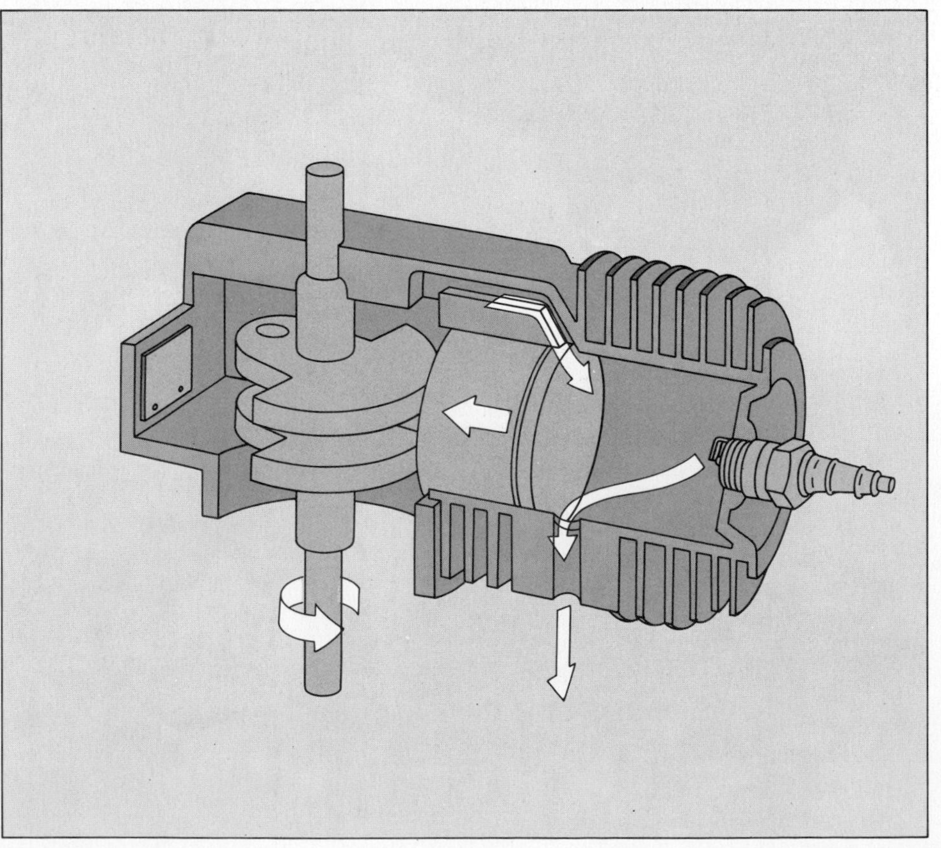

Keeping an Engine Oiled and Cool

The interior of a small gasoline engine is a miniature inferno, every part of which is subjected to extreme stresses of friction and heat. Thousands of times a minute, metal piston rings rub up and down against the cylinder wall. Combustion temperatures soar above 2,000° F. Such brutal conditions, if not moderated, would soon torture any metal to scrap. What enables an engine to endure these constant rigors are two vital systems: cooling and lubrication.

The worst of the searing heat, a by-product of the repeated explosions of atomized gasoline and air, is generated in the combustion chamber. Roughly one third of the heat is exhausted with the spent fuel vapor, while another third is consumed productively in driving the crankshaft. Although it is only a final third of the total, the remaining heat must be dissipated, because it is still far above the temperature that the engine metal can withstand.

To disperse this last residue of heat, the engine gulps in huge volumes of air. The cooling streams are wafted around the cylinder by a fan—actually the blades of the spinning flywheel *(right)*. The air shroud, a contoured duct housing the flywheel, directs the flow around the extruded cooling fins on the cylinder, which increase the surface area of the engine block and permit heat to escape more rapidly.

Friction, no less a threat to engine life than heat, results inevitably as parts slide on, over or through one another. This friction can both scratch vital parts and generate still more heat, which in turn expands the moving parts and creates even more friction.

The rapid, irreversible damage that this chafing could cause is prevented by the lubrication system, which distributes oil to all moving engine parts. In some four-cycle engines, ladle-like devices, bolted or cast onto the connecting rod and descriptively called dippers or splash fingers, dredge oil from the crankcase and splash it onto critical parts with each revolution of the crankshaft. In other engines, a tiny pump squirts oil through tubes in the crankshaft.

In two-cycle engines, oil is mixed with the fuel and thus is carried directly into the cylinder. After it has served its lubricating purpose, it is burned along with vaporized gasoline and air during the combustion cycle. This method of lubrication permits two-stroke engines to function normally at any angle, making them ideally suited for tools such as chain saws and weed trimmers, which are often tilted every which way during use. Four-stroke engines must remain upright at all times lest oil spill from the crankcase reservoir.

The type of oil used is also vitally important to the survival of the engine. Two-stroke engines, for example, cannot tolerate automobile engine oil, which breaks down under the normally higher operating temperatures. It also contains detergent additives that would foul spark plugs and clog exhaust ports with unburned residues. Use only specially labeled two-stroke engine oil and, when you buy it, check to see if it calls for a particular gasoline: Some two-stroke engine oils require leaded regular gasoline; unleaded grades sometimes cause an engine to run hotter.

Four-stroke engines, too, may need special lubrication. Engines used in winter, such as those on snow blowers, start more readily with multiviscosity oils, which become thinner as temperatures drop instead of thickening in frigid weather as ordinary oils do.

Cooling and lubricating an engine. Cooling air, drawn into the engine by the spinning flywheel, is directed to the cooling fins of the cylinder by the sheet-metal air shroud; the intake screen of the shroud prevents leaves and other debris from being drawn into the engine. Meanwhile, with each revolution of the crankshaft a splash finger on the connecting rod dips into the crankcase oil reservoir and flings oil over the engine's moving parts.

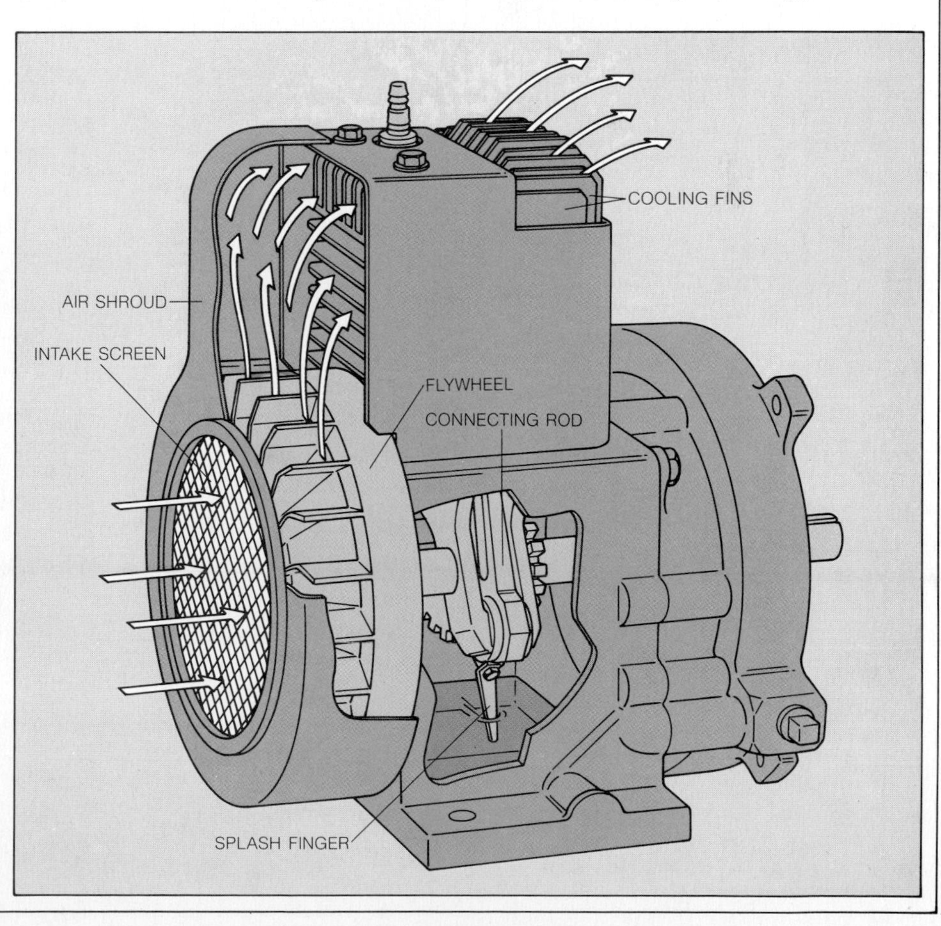

Commonsense Care to Keep an Engine Running

Small gasoline engines have earned such a reputation for ruggedness and reliability over the years that many users, taking them for granted, tend to neglect basic maintenance. But even the toughest engines need regular, commonsense care.

Maintenance routines vary with the engine, of course, depending on its size and purpose as well as on the frequency and even the season of its use. Snowblowers, for example, do their work in cold, generally damp weather and require different care from mowers and tillers, which operate in hot, dusty conditions. And maintenance for a mower that is used 10 months of the year in the southern United States is different from that recommended for the same machine when it is used for only three months in the northern United States or Canada.

The key to proper engine maintenance lies in frequent reference to the owner's manual, which lists specific requirements, procedures and time intervals. Following are general recommendations for handling fuel and oil and for cleaning the engine and its various filters.

Fuel

Fill the fuel tank of a four-stroke engine with regular-grade, leaded gasoline unless the manufacturer specifies another type. Do not use additives, which can leave residue in the tank and damage carburetor seals. For two-stroke engines, use the fuel-oil mixture recommended by the manufacturer of the engine. If possible, fill the fuel tank while the engine is cold, and make sure the gas-cap vents are not clogged.

Do not try to store gasoline for periods longer than four months; it will evaporate, leaving deposits of gummy sediment in its container.

Oil

Check the oil level in the oil reservoir of a four-stroke engine each time you use the engine; some reservoirs have dipsticks, but on others, you have to look directly into the filler tube. Add oil whenever necessary to bring it to the proper level. After about 25 hours of operation—or about 12 hours in dirty or dusty operating conditions—drain and replace the oil while the engine is warm.

Always use the type of oil recommended by the manufacturer. Four-stroke engines use automotive oil. During the summertime and in temperatures higher than 40° F., a standard SAE 30-grade oil is generally recommended. When temperatures are lower, a lighter, smoother-flowing SAE 5W-20- or 5W-30-grade oil is preferable. The oil used in two-stroke engines is combined with gasoline and added directly to the fuel tank *(box, page 30)*. Unlike automotive oil, this special, additive-free oil is highly combustible; be sure to consult the owner's manual for the type and quantity recommended by the manufacturer.

Fuel Filters

Filters for two-stroke engines are usually located at the end of the fuel line, inside the fuel tank. Some can be lifted out of the tank's filler opening for cleaning *(page 31)*; others cannot be cleaned and must be replaced. Check for instructions in the owner's manual.

On larger, four-stroke engines you may find a detachable fuel filter inside the tank or on the line connecting the tank to the carburetor. The metal or fiberglass mesh strainers and the plastic or glass sediment bowls used for these filters ought to be taken off, soaked in kerosene, rinsed and dried after approximately 25 hours of engine operation. There is a fuel shutoff valve so that you can stop the flow of gasoline before you remove the filter.

Air Cleaners

All two- and four-stroke engines have air cleaners that filter air through oil, oiled foam, mesh or paper to remove impurities before the air enters the carburetor. The three most common types of cleaners are shown on pages 28-29; each should be cleaned by the method described, with the frequency recommended by the manufacturer.

Exteriors

All engines become dirty with use and must be cleaned regularly; heavy coats of grease and grime not only keep an engine from working at peak efficiency but can also cause permanent damage from overheating. These deposits can be removed with a spray solvent, available at hardware stores; follow the method shown on page 30.

Two-stroke engines are plagued by carbon build-up, which results from incomplete combustion of the oil and gasoline that are combined in the fuel tank. The carbon accumulates on the muffler and exhaust ports and should be scraped off regularly, as shown on page 31. A cleaning after every 25 hours of operation is generally recommended; chain saws may need more frequent cleaning because they run at full throttle most of the time.

Tips for Safety on the Job

When undertaking any maintenance or repairs on a small engine, keep in mind that you are dealing with a machine that contains highly flammable fuel and that produces a pair of potentially dangerous by-products—heat and exhaust fumes. You can protect yourself by observing the following safety precautions:

□ Allow a hot engine to cool for at least 30 minutes before handling or refueling.
□ Before you begin work on an engine or its attachments, always set the throttle to STOP and disconnect the spark-plug wire to prevent accidental starting.
□ Never smoke while refueling.
□ Do not operate or refuel an engine indoors without adequate ventilation.
□ Do not store quantities of fuel. Keep only small amounts in vented gasoline containers, away from heat and from direct sunlight.

Two Ways to Drain Oil from a Four-stroke Engine

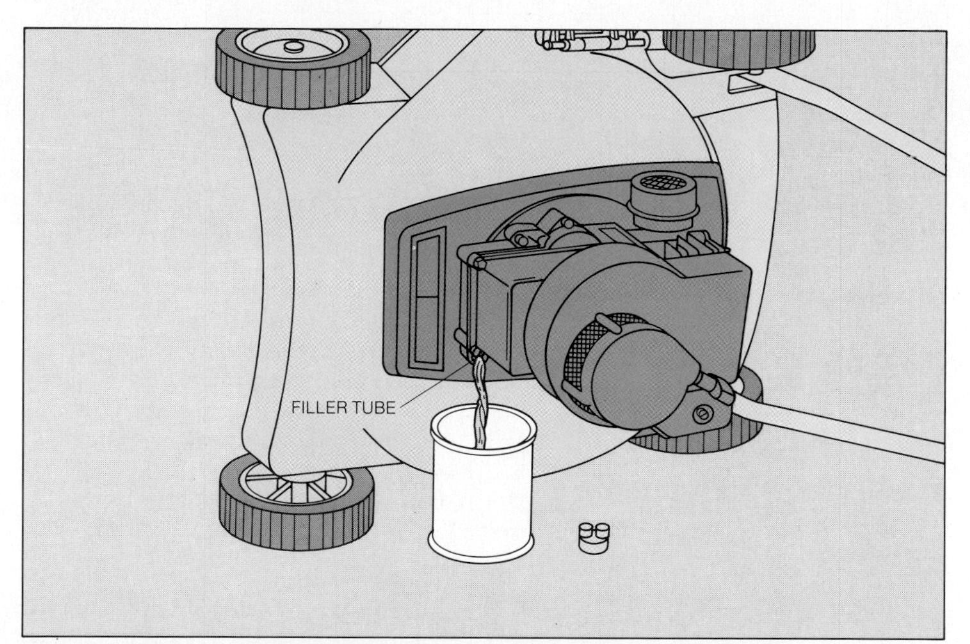

Draining oil through the filler tube. After draining the fuel tank *(page 54)*, uncap the oil filler tube and slowly tilt the mower so that oil flows from the mouth of the tube into a wide-necked container—a coffee can is a good disposable oil collector. Leave the mower resting on its side—the push handle will act as a support—until all of the oil has drained out of the tube. If the oil does not flow freely, tilt the mower farther toward the container and prop it in that position with a few scrap boards or bricks.

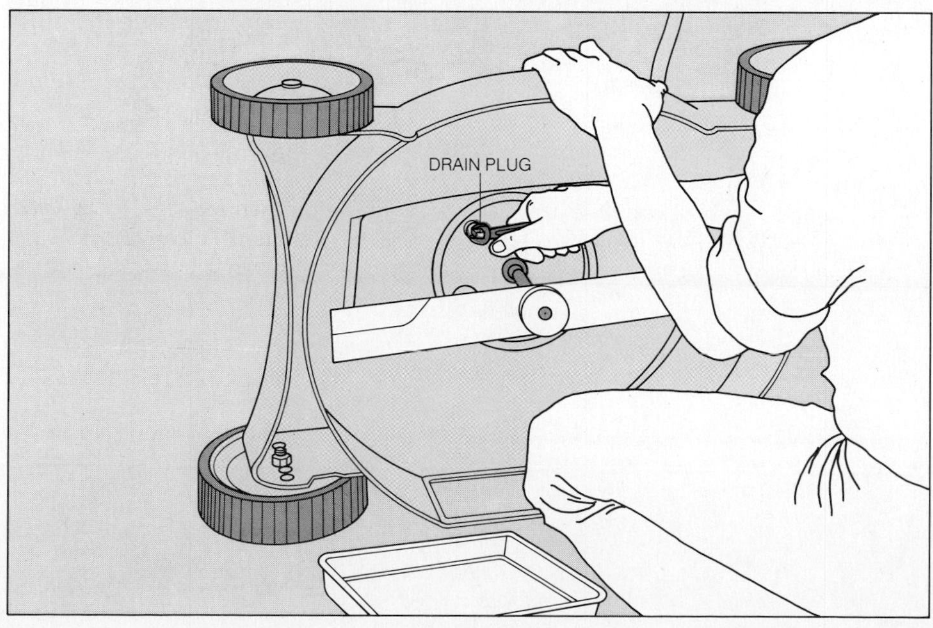

Emptying an oil sump. Drain the fuel tank, then tilt the mower on its side so that the oil-sump drain plug on the underside of the engine is above the crankshaft. Remove the drain plug *(left)* and return the mower to an upright position with the drain opening directly over a shallow pan. If the oil drains too slowly, raise or prop the wheels about 1 inch on the side of the mower opposite the sump drain.

Washing an Oiled-foam Air Cleaner

1 **Removing the old oil.** To wash the oiled-foam cleaner, which screens air entering the carburetor by passing it through a filter of oiled polyurethane foam, remove the central setscrew *(inset)* that secures the cleaner to the mouth of the carburetor and lift the cleaner off the carburetor. Snap the cover off the rim of the housing and remove the metal spacer and the foam filter. Then wash the foam in a bucket of kerosene, soaking and squeezing it repeatedly until all of the oil is washed out. Dry the foam by wrapping it in paper towels and squeezing it. Then wash the remaining parts of the cleaner, rinse them in water and wipe them dry.

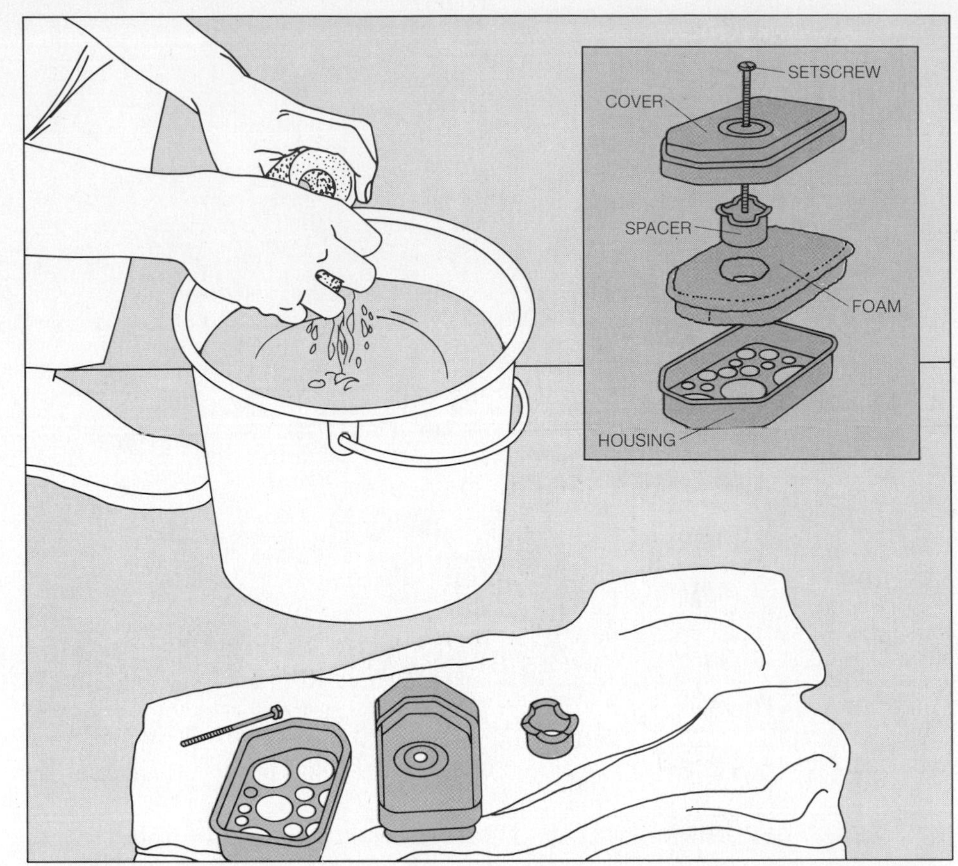

2 **Oiling the clean filter.** Drip clean engine oil onto the foam filter, squeezing the foam to distribute the oil evenly. When the foam is saturated, squeeze it once more to remove excess oil and then set the filter back in the housing. Be sure that the top edge of the filter is resting on the lip of the housing *(inset, top)* to form a continuous seal; then replace the spacer, snap on the housing cover and reattach the filter to the carburetor with the central setscrew.

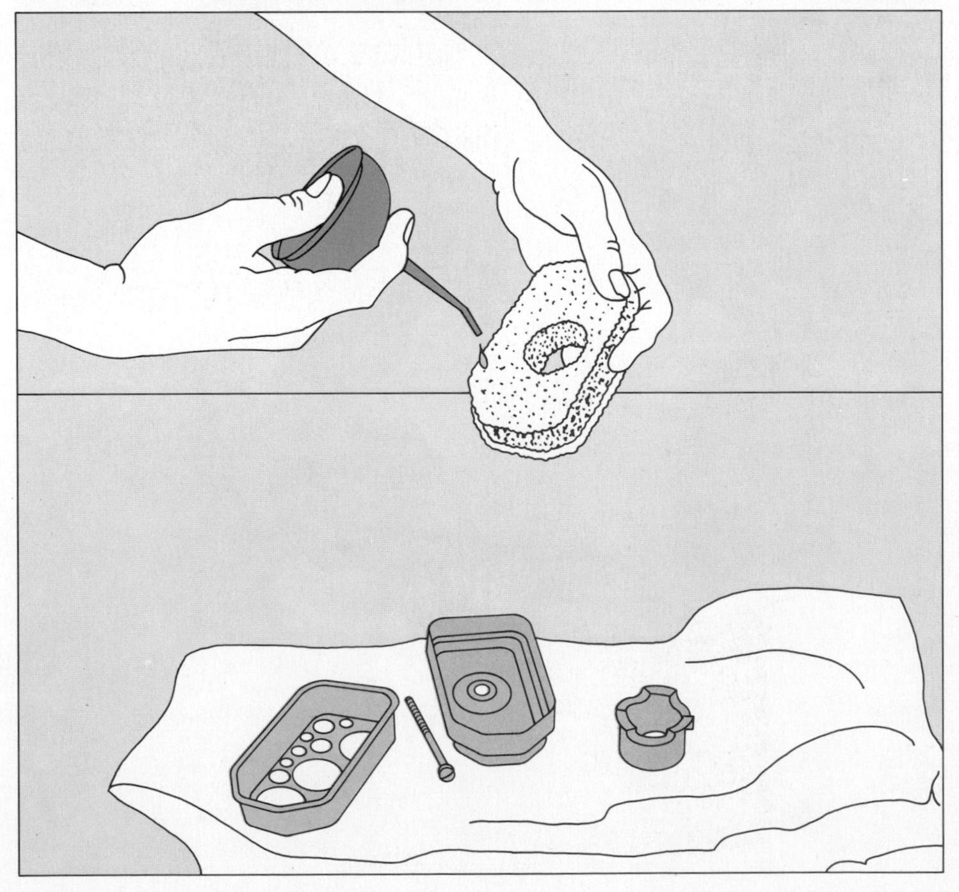

Alternative Types of Air Cleaners

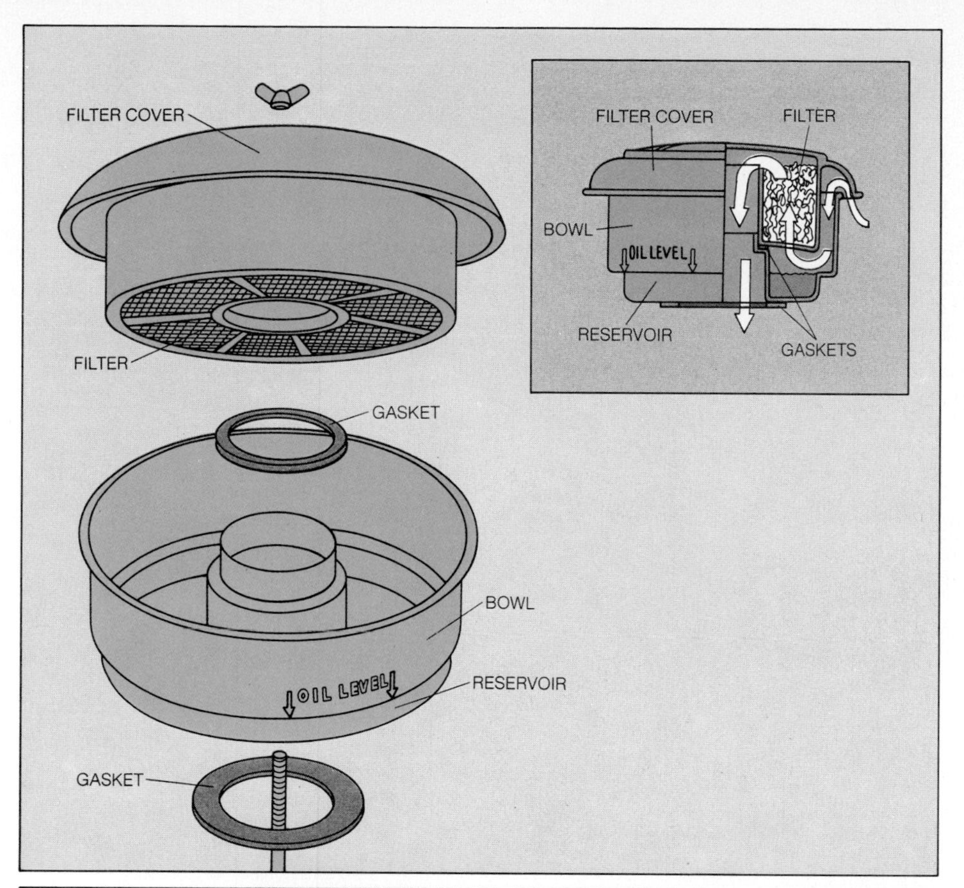

An oil-bath cleaner. Often used for engines that operate in very dusty conditions, the oil-bath cleaner directs incoming air up under the edge of its filter cover *(inset),* down into a filter bowl and over a shallow reservoir of oil that moistens the air and removes some of the larger dust particles. The air then flows up through a mesh or foam filter and back down through the center of the cleaner to the carburetor.

To wash an oil-bath cleaner, unscrew it from the carburetor, lift the filter out of the bowl and pour the oil out of the reservoir. Wash the bowl and filter with kerosene. Pour clean engine oil into the reservoir up to the oil-level indicator line. If the gaskets are worn or damaged, replace them.

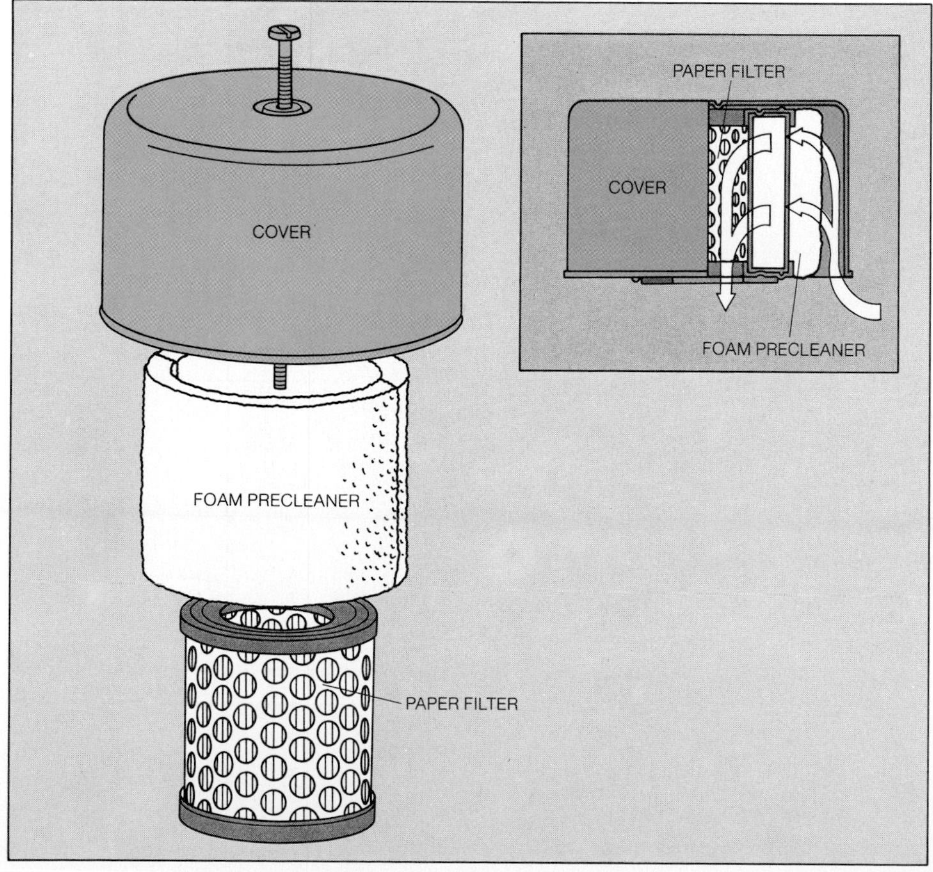

A dry-filter cleaner. The dry-filter cleaner—a miniature version of the air filter used in many automobiles—directs incoming air up under its cover, through an oiled-foam precleaner, through a paper filter treated with a moisture-resistant chemical, then down into the carburetor *(inset).* The oiled-foam precleaner is washed and re-oiled in the same manner as the oiled-foam cleaner *(opposite).* Remove dirt particles from the paper filter by tapping it on a hard surface. If the filter is badly clogged, replace it.

Degreasing a Four-stroke Engine

1 Protecting the vital parts. Start the engine and allow it to run for a couple of minutes, then turn it off and remove the engine shroud and the air cleaner. Cover the fuel cap, the mouth of the carburetor and the spark plug with pieces of plastic wrap or aluminum foil.

2 Spraying grimy surfaces. Wearing a respirator and goggles, spray or brush engine-degreasing solvent liberally over greasy engine surfaces and between the cooling fins. Wait for 10 to 15 minutes—as directed on the solvent can—then hose the solvent off with water. If necessary, scrub between the fins with a soft brush and more solvent, then rinse again. Remove the plastic or foil, replace the shroud and air cleaner and run the engine for 15 minutes to dry it.

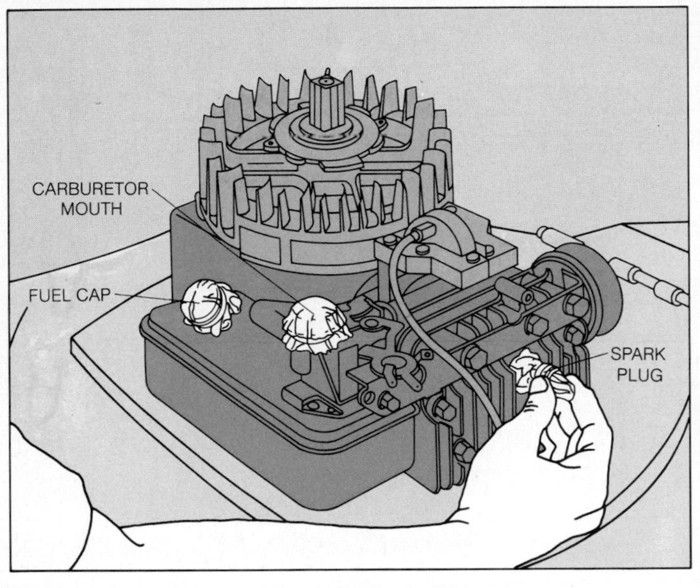

Blending Gas and Oil for a Two-stroke Engine

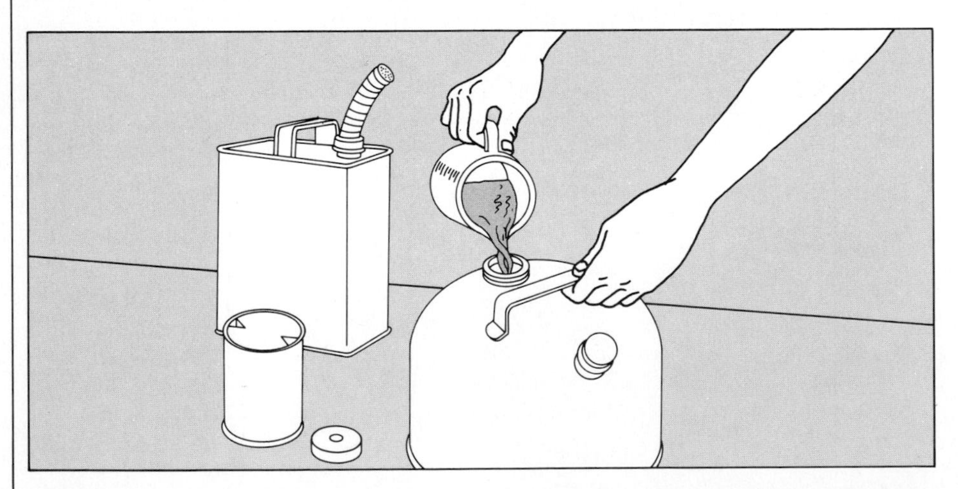

Unlike four-stroke engines, where the gasoline that fuels the engine and the oil that lubricates it are separate, two-stroke engines use a premixed combination of gasoline and oil. Mixing the two is a simple matter, once you have checked the owner's manual to make certain that you are using the correct ingredients in the specified proportions.

Most manufacturers recommend the use of regular-grade, leaded gasoline and a special blend of oil, sometimes called outboard oil, that contains no detergents or other noncombustible additives. One engine may require 8 parts gasoline to 1 part oil, while another may call for a 50-to-1 mixture. Inaccurate measurement will cause engine problems: Too little oil can cause overheating; too much oil will produce smoky exhaust, fouled plugs and misfiring.

It is best to prepare no more than a gallon of gasoline-oil mixture at a time, because gasoline evaporates far more rapidly than oil and, over a period of time, this evaporation would alter the prescribed ratio.

To mix the two ingredients, pour half of the gasoline into a closed container and add all of the oil; use a measuring cup if necessary. There are 128 fluid ounces in a U.S. gallon and 160 ounces in a Canadian gallon, so oil measurements will vary from about 2½ ounces per U.S. gallon of gasoline (3 ounces per Canadian gallon) for a 50-to-1 ratio to 16 ounces (20 Canadian ounces) for an 8-to-1 ratio. Shake the partial mixture vigorously, then add the remaining gasoline and shake the mixture again.

How to Clean Two-stroke Engines

Scraping the muffler. Remove and disassemble the muffler and scrape encrusted carbon from all surfaces, using a metal blade *(below, left)* or a stiff brush. Turn the engine flywheel by slowly pulling the starter until the piston covers the exhaust ports, then use a wooden scraper to clean carbon from around the ports *(below, right);* the piston will keep debris from falling into the cylinder. Do not scratch the piston. Blow away any loose carbon, then replace the muffler.

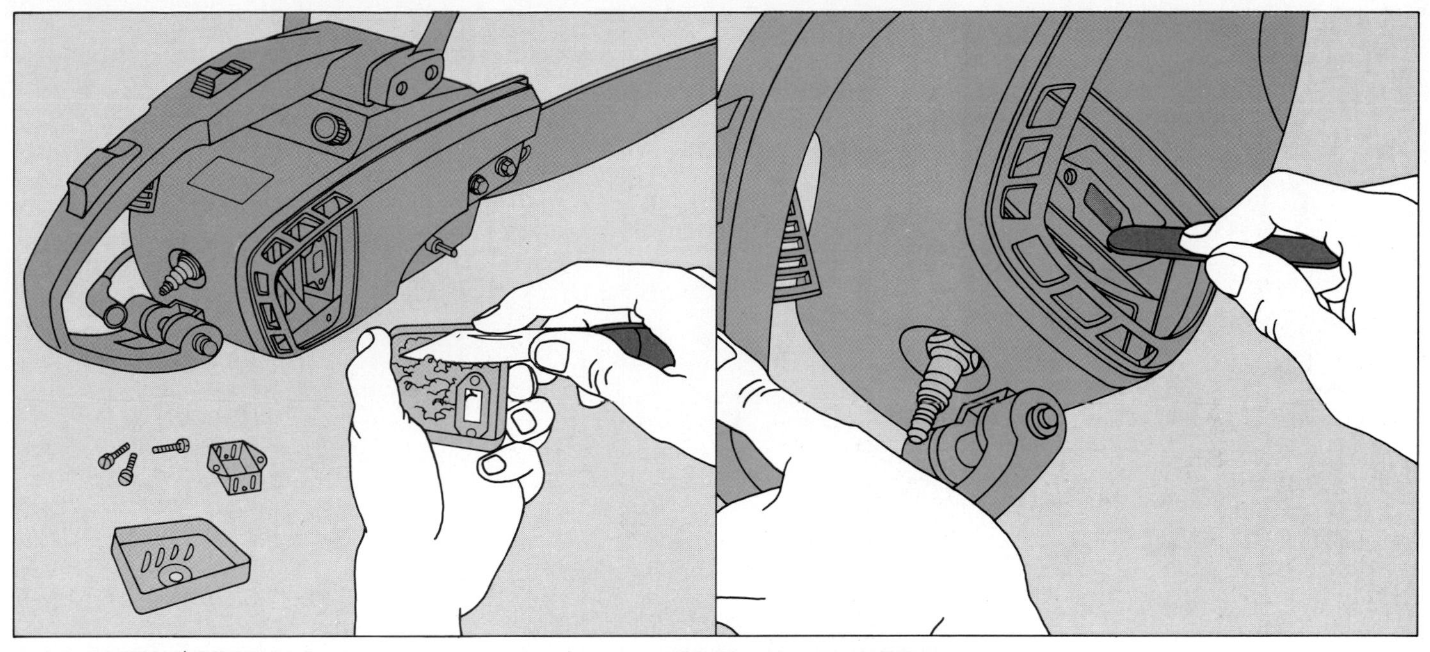

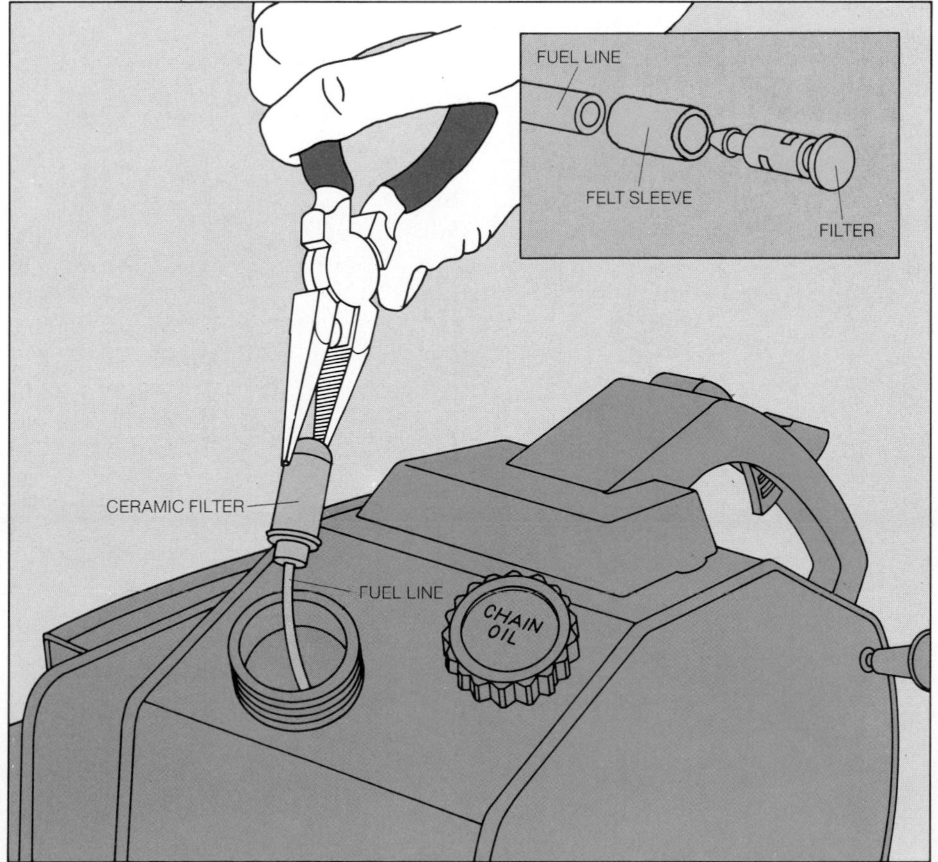

Cleaning the fuel filter. Pour the fuel out of the tank, lift out the fuel line and filter with long-nose pliers, and twist the filter off the line. It may be necessary to remove the fuel line from the outside of some tanks. Clean a ceramic filter with kerosene and dry it before replacing it. If the filter assembly includes a removable felt sleeve *(inset),* wash it with kerosene, rinse with water and dry thoroughly.

Troubleshooting an Engine for Simple Repairs

When a small engine is ailing, one of the first symptoms it is likely to display is an adamant refusal to start. In most cases, simple first aid will bring it coughing back to life. Sometimes, however, a balky engine requires major surgery. The first step in determining which remedy to apply is to narrow the problem down into one of three categories.

In order to run, a gasoline engine needs a strong spark, a proper fuel mixture and sufficient compression. When it will not start, one of these vital elements of combustion is almost certainly missing. You can easily determine which of the three is the basic trouble spot with a simple four-step test *(opposite)* that takes only minutes to perform and requires the removal of only one engine part—the spark plug. Without disassembling the engine, you can eliminate whole engine systems from suspicion on by one, and simplify your search for a specific failing component by drastically reducing the number of suspect parts.

Before running the test, rule out the most obvious possibilities. First, fill the gas tank with fresh gasoline. Old gas may not burn, and fuel from an old storage container may contain rust particles that can clog the carburetor. Next, set the throttle to the choke position; and if the engine has an automatic turn-off feature, such as a fuel shut-off device, make sure it is disengaged.

If you have made several attempts to start the engine and can smell a strong odor of gasoline, the engine probably is flooded. The excess gasoline must be evacuated from the cylinder *(page 35)*; otherwise, additional attempts to start the engine will only worsen the flooding. After the flooding has been corrected, the engine may start. It should be checked out nonetheless for ignition or compression problems that may have prevented it from starting—and caused the flooding—in the first place.

The lack of a spark to ignite the fuel in the combustion chamber is one of the most common reasons an engine will not start. To produce a spark, the ignition system generates high-voltage bursts of electricity that reach the spark plug through the spark-plug wire. You can judge the ignition system visually by disconnecting the wire from the plug and pulling the starter cord. If the wire sparks *(opposite, top left)*, indicating that the ignition system's output is sufficient, the spark plug itself may be at fault; it should be removed and checked *(opposite, bottom left)* to rule out ignition problems.

As soon as you remove the plug, you will see evidence of any fuel-flow problems the engine may have: A dry tip is a sure indication that no fuel is reaching the cylinder. The plug may also give you valuable clues about problems elsewhere in the engine. Constantly exposed to combustion, used plugs bear telltale signs of engine performance in the color and texture of the deposits that accumulate on them *(page 34)*.

If your inspection of the plug has shown that no fuel is reaching the cylinder, the cause is most likely a blockage or a stuck mechanical part. Some blockages are easy to reach and fix. Most often overlooked are simple but disabling clogs in the gas cap: When the tiny vent holes that allow gasoline to flow freely from the tank to the carburetor become clogged, a vacuum forms in the fuel line and the flow stops, causing the engine to turn off intermittently. The holes are easily cleaned out with a pin or a wire *(page 35)*. On the other hand, a clogged air filter can produce flooding by preventing air from flowing to the carburetor. If the foam filter is covered with grime when you remove its housing from the carburetor, it will have to be either cleaned or replaced *(page 28)*.

Other fuel blockages are less accessible and may be less obvious. Blockages that form in the tiny chambers of the carburetor and fuel line are among the most difficult to find. Diagnosing them involves dismantling the fuel system *(pages 58-69)*. Pinning down mechanical malfunctions of the delicate carburetor linkages also requires disassembly.

Much easier to detect is a special fuel-flow problem that occurs when the intake valve of a four-stroke engine or the reed valve of a two-stroke engine becomes stuck open: The air-fuel flow reverses direction, being forced back through the carburetor through the open valve. Although valves are among the most inaccessible of all engine parts, this valve problem can be diagnosed from outside the engine simply by observing the carburetor while the starter cord is pulled *(page 36)*.

In rare instances, an exhaust blockage may keep an engine from starting. A clog in the exhaust port or the muffler can inhibit the outrush of burned fuel from the cylinder and obstruct the inflow of fresh fuel. The exhaust port of a small gasoline engine is typically only about an inch long and very easy to inspect: Simply unscrew the muffler from the engine block *(page 36)*.

A severe loss of compression—which is caused by a leaky cylinder—can be detected by the finger test shown opposite, bottom right. Compression loss—except in the case of a loose spark plug—is invariably a serious matter and requires the major repairs shown on pages 70-83. First, however, the area of the leak should be identified by a second test *(page 37)*, in which oil is squirted into the cylinder to determine whether the leak is caused by the piston or piston rings, the valves or the head gasket.

Electric starting systems on the larger engines that have them are prone to special problems. When the engine will not turn over, the battery is the first part to test, using an inexpensive hydrometer that resembles a large eyedropper *(page 37)*. If the battery is fully charged, examine the starter motor *(page 90)*.

Finding the Basic Faults

1 Testing the ignition system. Pull the spark-plug wire off the spark-plug terminal. Wearing a glove to guard against shock, hold the wire ¼ inch from the terminal with one hand and pull the starter cord with the other hand. A blue spark should jump from the wire to the terminal, making a snapping sound. If there is no spark, the ignition system is malfunctioning. Repair it as shown on page 38.

In bright sunshine it may be difficult to see the spark. To make it visible, perform the test in a darkened area such as a garage or a workshop. If you find it hard to pull the cord and watch for the spark at the same time, enlist a helper.

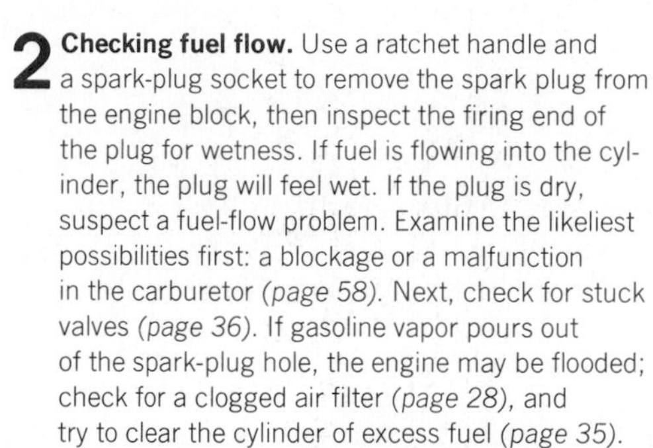

2 Checking fuel flow. Use a ratchet handle and a spark-plug socket to remove the spark plug from the engine block, then inspect the firing end of the plug for wetness. If fuel is flowing into the cylinder, the plug will feel wet. If the plug is dry, suspect a fuel-flow problem. Examine the likeliest possibilities first: a blockage or a malfunction in the carburetor *(page 58)*. Next, check for stuck valves *(page 36)*. If gasoline vapor pours out of the spark-plug hole, the engine may be flooded; check for a clogged air filter *(page 28)*, and try to clear the cylinder of excess fuel *(page 35)*.

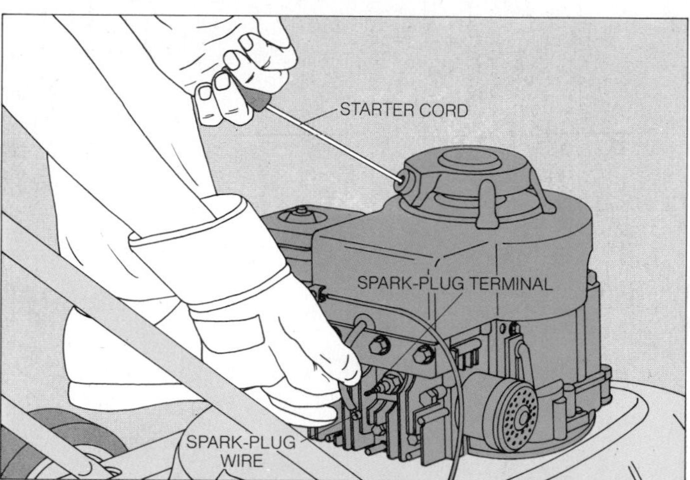

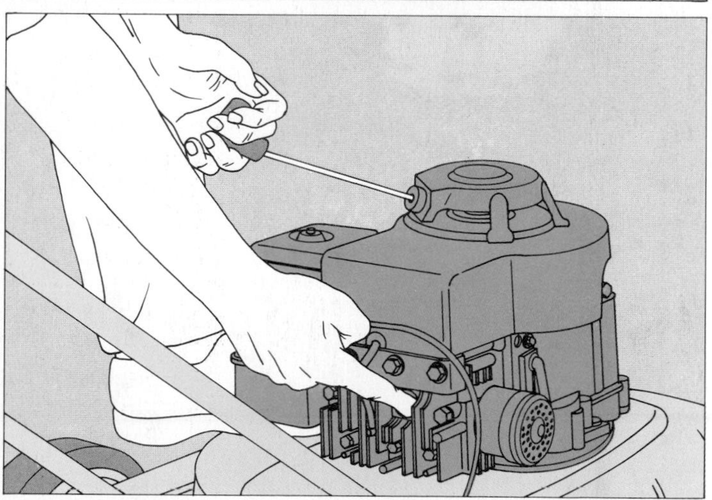

3 Testing the spark plug. Examine the tip of the spark plug for oil, soot or particle deposits, then wipe it clean and reconnect the spark-plug wire to its terminal. Touch the plug's metal threads against the bare metal of the engine block—away from the spark-plug hole—then pull the starter cord and watch for a spark between the plug electrodes. If there is no spark, the plug is defective. Replace it with a new one.

4 Checking compression. Hold a finger over, but not in, the spark-plug hole in the engine block and pull the starter cord. The air pressure inside the cylinder should alternately draw your finger toward the hole and blow it away. If you do not feel air pressure, lack of compression is preventing the engine from starting. A leaking head gasket or worn valves, piston or piston rings are likely culprits. Perform an oil test *(page 37)* to investigate further.

The Telltale Condition of a Spark Plug

A normal plug. Slightly rounded electrodes with a light coat of ashy brown to grayish tan deposits indicate normal wear. They are evidence of an engine in good condition and a carburetor in correct adjustment.

Oil fouling. A wet, oily film at the tip of a spark plug will prevent sparking. In a two-stroke engine, spraying aerosol spark-plug cleaner on the electrodes and wiping them dry will usually solve the problem quickly—oil fouling is common in two-stroke engines and seldom means serious trouble. However, in four-stroke engines the condition signals worn valves or piston rings; both require an overhaul *(pages 70-83)*.

Carbon fouling. Dry black soot covering the spark-plug tip indicates that the fuel is not burning completely. The problem is caused by a too-rich fuel mixture (an oversupply of gasoline in relation to air), a weak ignition or both. To correct the fuel mixture, clean the air filter *(page 28)* and, if necessary, adjust the carburetor *(page 53)*. To improve the ignition, perform a tune-up on the breaker points *(pages 38-46)*.

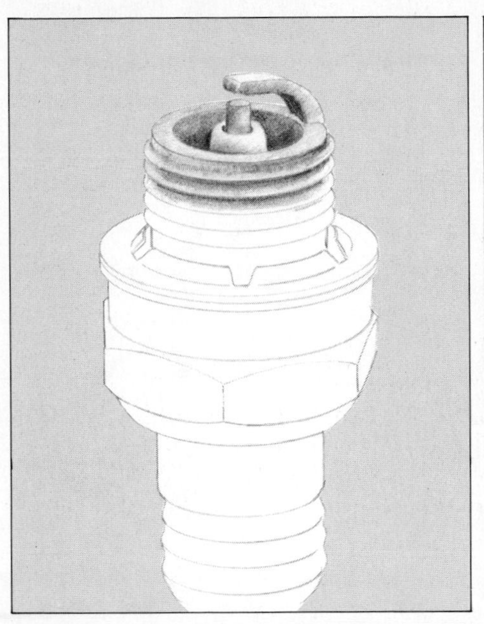

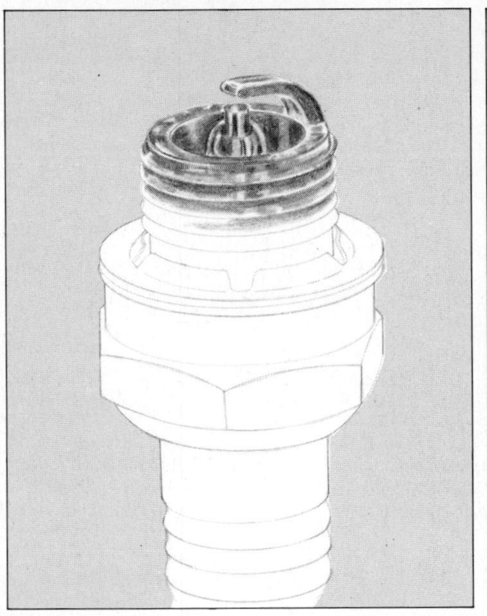

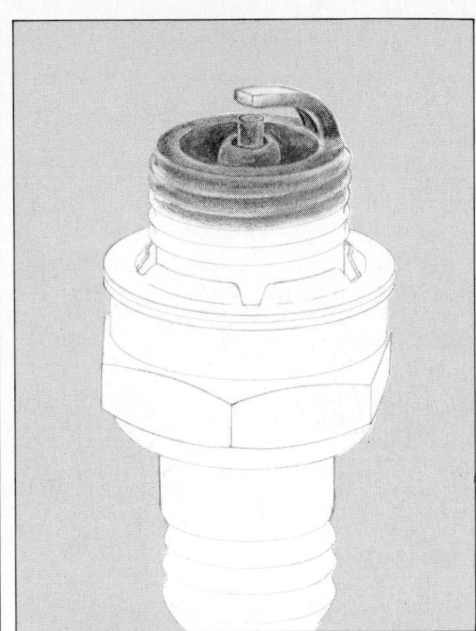

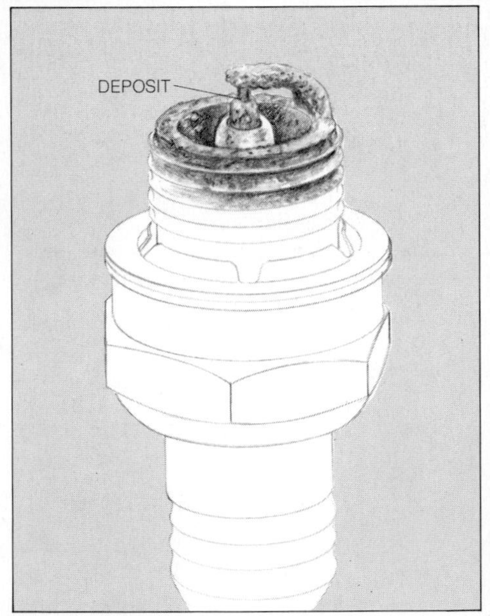

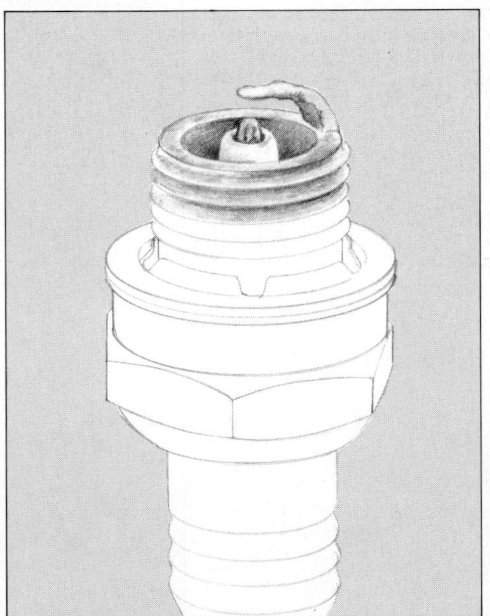

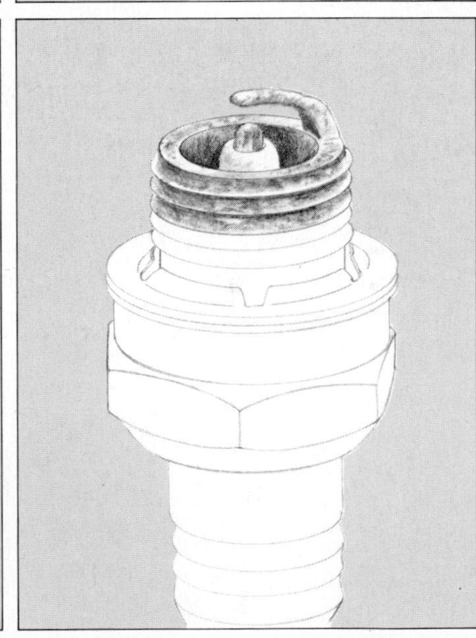

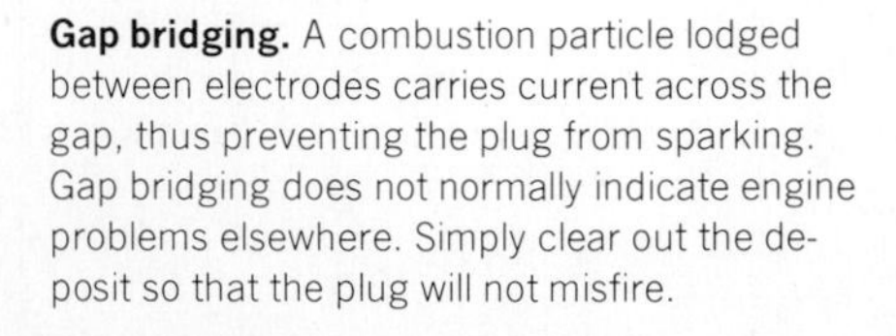
Gap bridging. A combustion particle lodged between electrodes carries current across the gap, thus preventing the plug from sparking. Gap bridging does not normally indicate engine problems elsewhere. Simply clear out the deposit so that the plug will not misfire.

Overheating. Whitened, eroded electrodes are evidence of combustion that is too hot and too rapid. The cause is probably a too-lean fuel mixture—not enough gasoline in relation to air—or incorrect ignition timing. To balance the mixture, adjust the carburetor *(page 53)*. To set the timing, do a tune-up *(pages 38-52)*. An incorrect replacement plug can cause overheating: Make sure the plug is the recommended type.

Excessive wear. Worn, rounded electrodes encrusted with deposits are signs of a worn-out spark plug. Nothing is wrong with the engine, but the plug has outlived its useful life and should be replaced with a new one.

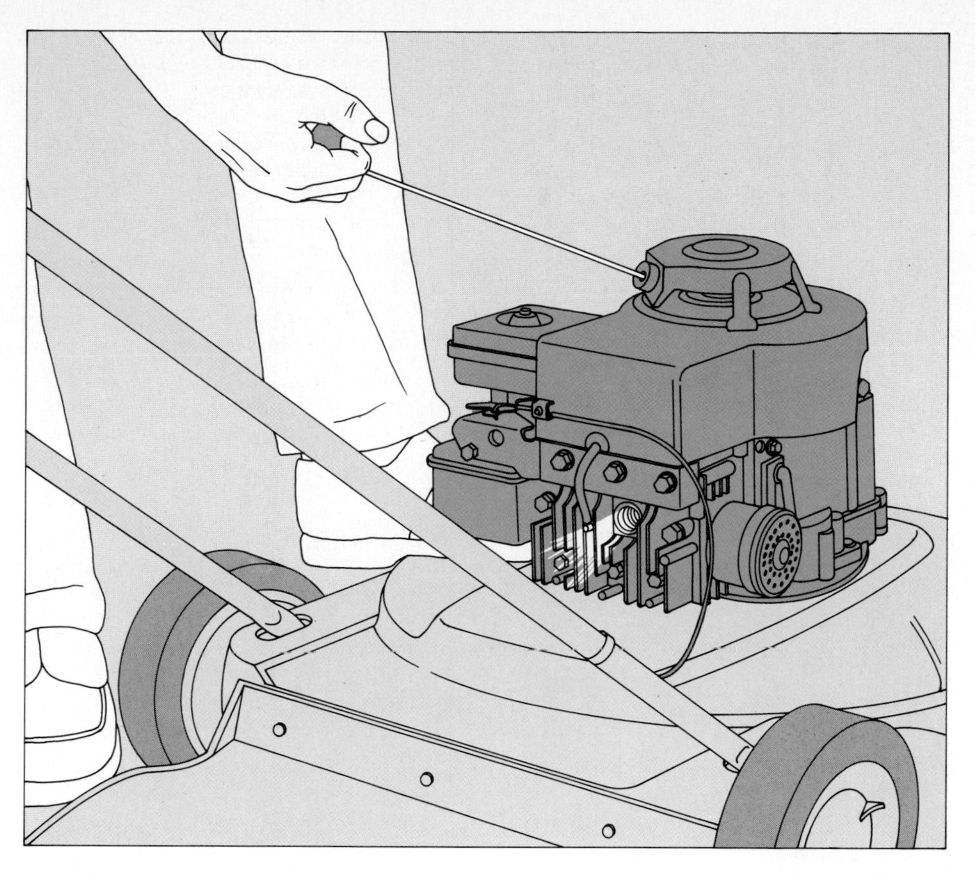

How to Start a Flooded Engine

Clearing a fuel-filled cylinder. To clear a flooded combustion chamber of excess gasoline, remove the spark plug with a ratchet handle and a spark-plug socket, turn the control lever to OFF and pull the starter cord five times. Screw the spark plug back in, turn the control lever to RUN and pull the starter cord again. The engine should start if flooding was the problem.

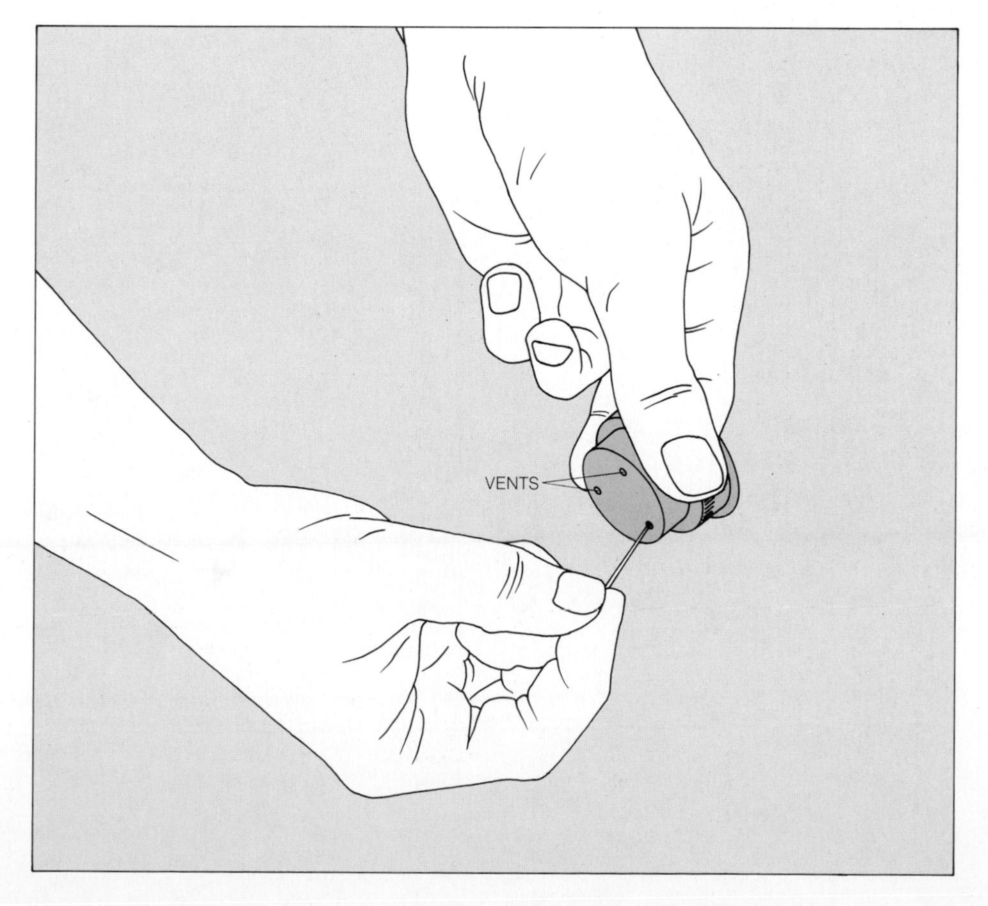

Dealing with Flaws in the Flow of Fuel

Unclogging a gas cap. If your engine has been stalling intermittently, remove the gas cap and try to start the engine. If the engine runs continuously only when the gas cap is off, the vents in the cap are clogged. To open them, probe through them with a pin or a wire, then screw the cap back on the gas tank.

Diagnosing a stuck intake or reed valve. Remove the air filter, place a hand close to the top of the carburetor and pull the starter cord. If you feel a blast of air, the engine has a serious valve problem. In a four-stroke engine *(inset, top)*, the rush of air is caused by an intake valve stuck in the open position. The open valve allows the air-fuel mixture—compressed by the ascending piston—to escape back through the carburetor *(arrow)*. Repair a stuck intake valve as shown on pages 70-71. The same symptom appears in a two-stroke engine *(inset, bottom)* when the reed valve does not close properly. When the piston descends, the air-fuel-oil mixture in the crankcase is forced back through the open valves and into the carburetor *(arrow)* instead of entering the combustion chamber. Repair a reed valve as shown on page 83.

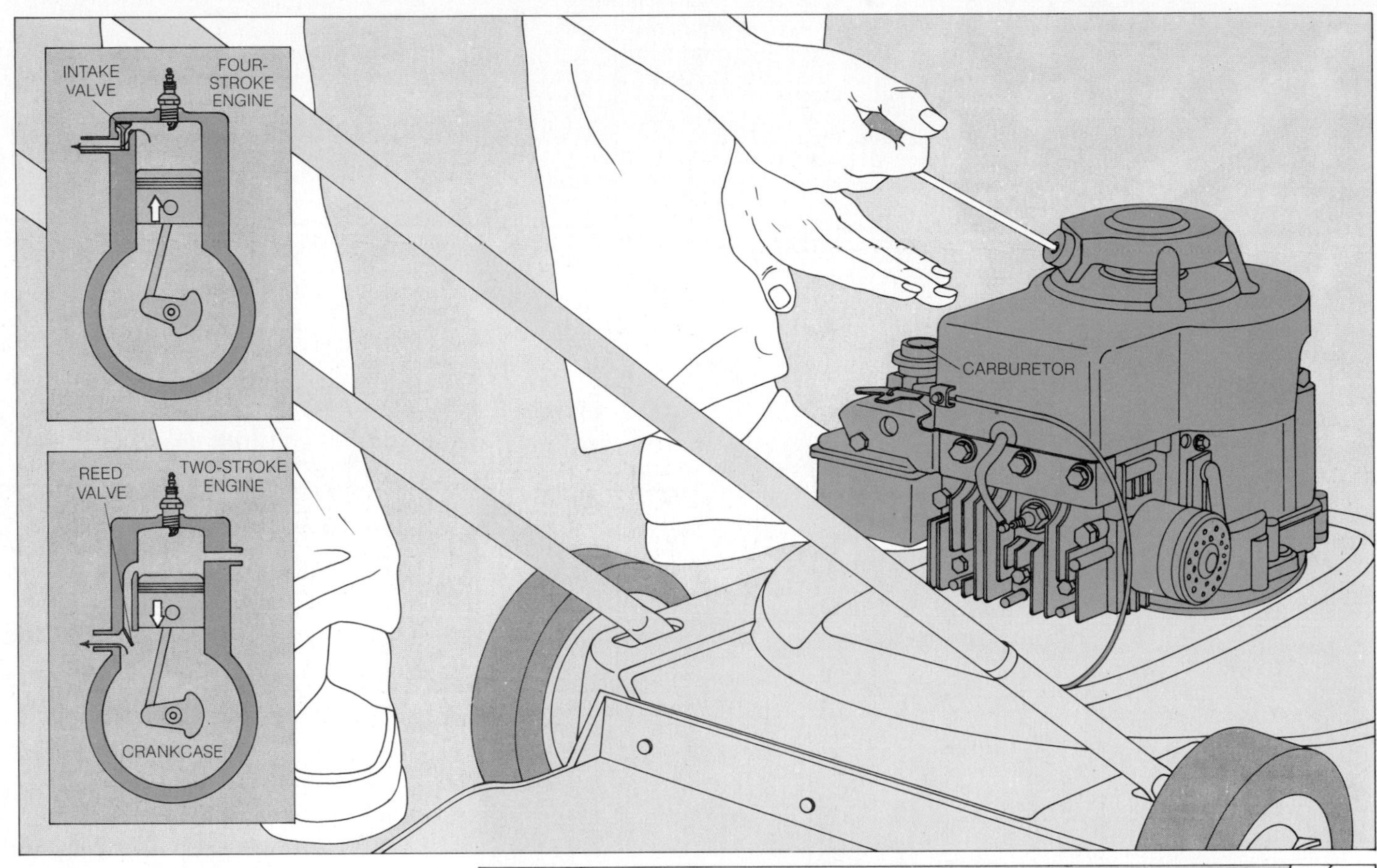

Finding and fixing exhaust blockages. Remove the muffler from the engine block. (On some engines, you will have to remove a long screw from the center of the muffler; on others, you can simply turn the muffler itself counterclockwise.) If the exhaust port is not located right next to the gas tank, you can try to start the engine without the muffler. If it starts, a clog in the muffler was the problem. Clean out any deposits from the exhaust port with a rag and replace the muffler with a new one.

If the muffler is located next to the gas tank, it is not safe to start the engine without it in place. Instead, install a new muffler before proceeding.

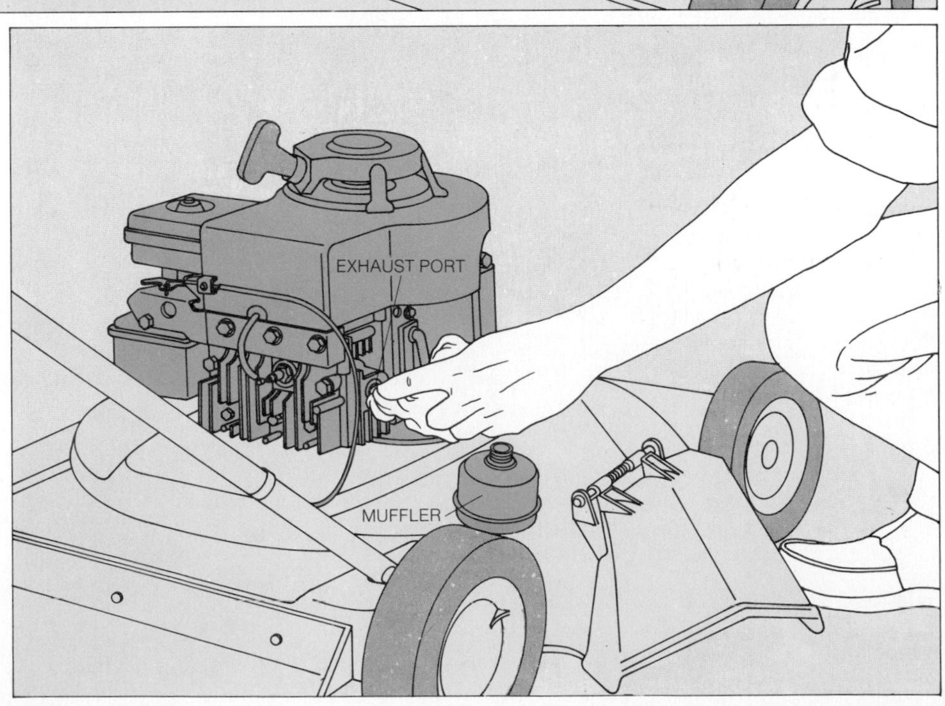

Diagnosing Compression Problems

Finding causes of low compression. Press the rubber end of a compression tester firmly into the spark-plug hole and give the starter cord a pull *(below, left)*. Remove the tester and note the reading on its dial. Then squirt an ounce of motor oil into the cylinder through the spark-plug hole *(below, right)*. Set the tester dial to 0 by pushing the reset button, then pull the starter cord and measure the compression again. If the second reading indicates increased compression, the piston rings or piston must be worn; the oil acted as a temporary seal between the rings and the cylinder wall. Repair the piston as shown on pages 70-83. If the reading for a two-stroke engine does not increase when oil is added, the cylinder head gasket is worn and leaking, causing compression loss. For a four-stroke engine, a leaking head gasket is one possibility; a stuck intake or exhaust valve may also be at fault. Replace gaskets or repair valves as shown on pages 70-71.

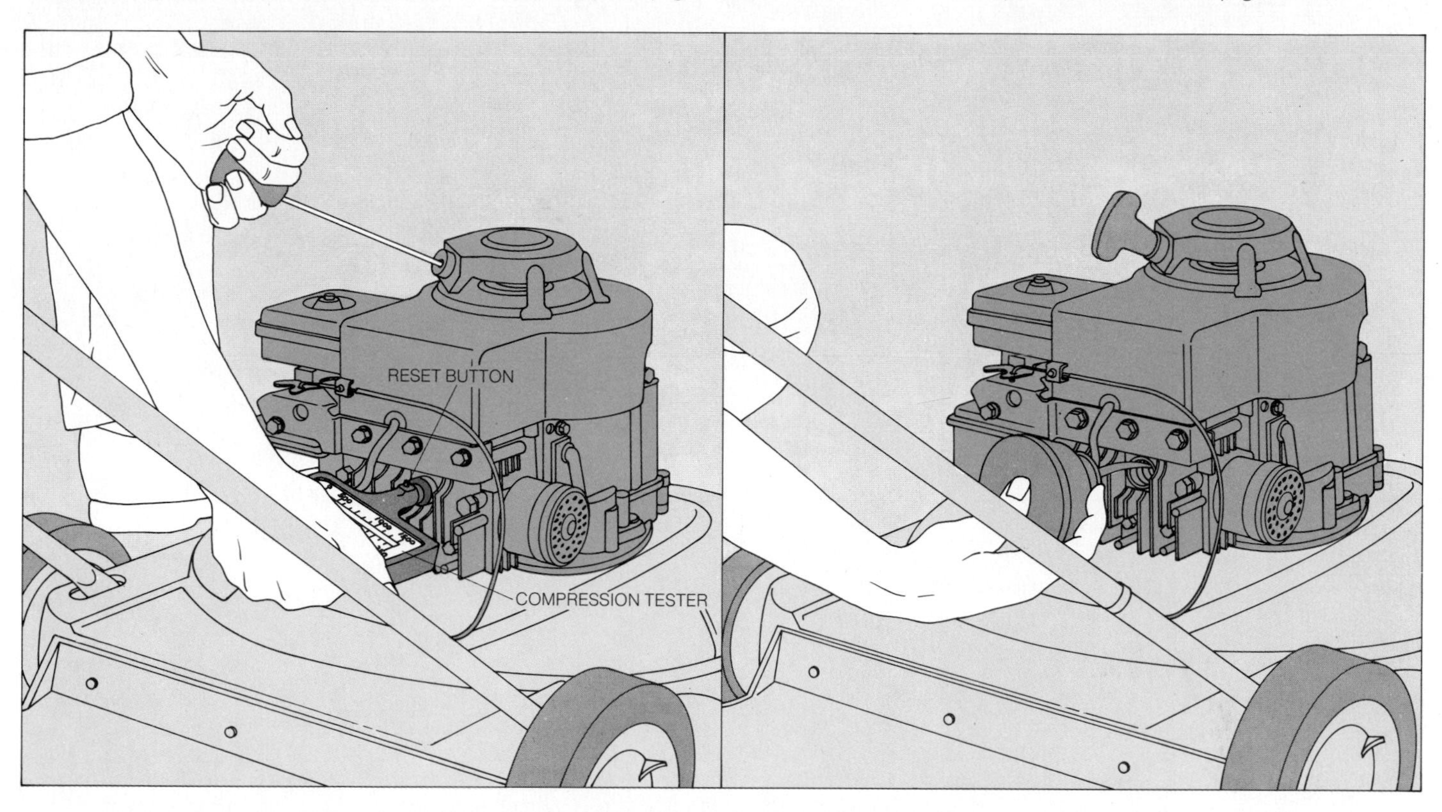

Checking the Battery of an Electric Starter

Testing battery strength. Remove the cap from a battery cell. Draw a sample of battery acid from the cell with a battery tester, and count the number of plastic balls that float to the top of the sample. To determine whether the cell is fully charged, check the number against the scale provided with the tester. Return the acid to the cell and similarly test all of the other cells.

If any of the cells are at less than full strength, have the battery charged by a small-engine repair shop. If the test indicates that the battery is fully charged, inspect and, if necessary, repair the starter motor *(pages 90-93)*.

The cells of a sealed maintenance-free battery are not accessible. Have the battery tested at a repair shop equipped with a device that can measure electrical load.

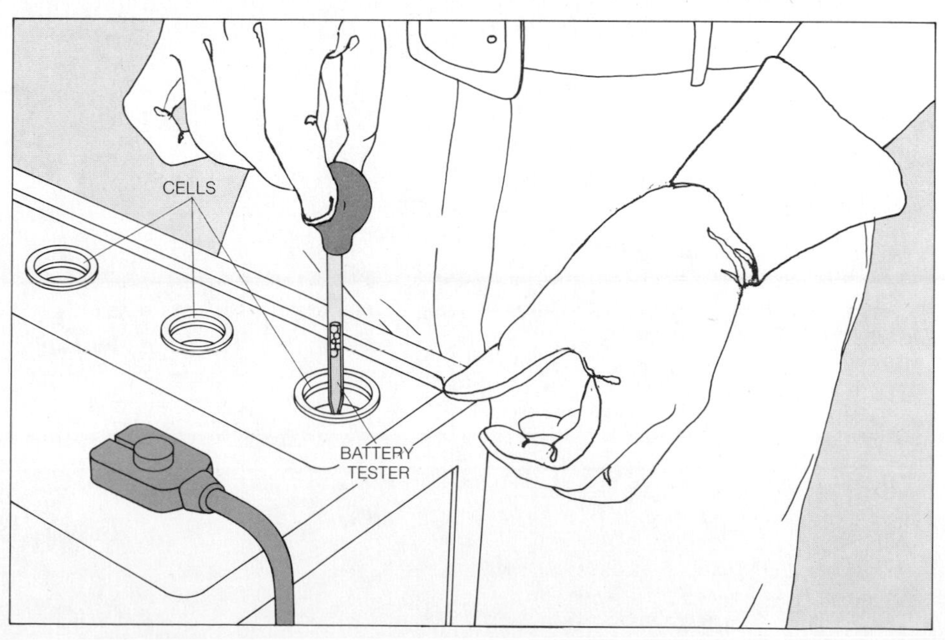

Beyond Simple Repairs: Tuning an Engine

A thorough tune-up will restore engine power and performance, which gradually diminish as the engine is used and parts begin to wear. But the procedure should not be postponed until the engine begins to sputter and stall. If done at proper intervals, tune-ups will reveal the symptoms of impending problems before the engine breaks down or requires major overhaul work.

The proper interval between tune-ups depends on how much use an engine gets. Naturally, a chain saw that is used for an hour every six months does not need the same yearly tune-up as a lawn mower that is used for several hours each week, but it may be wise to tune the saw before embarking on a major cutting project. As a rule of thumb, a small engine should get a complete tune-up after every 40 to 50 hours of operation, or before every new season of hard use.

Professional mechanics differ in their recommendations as to the scope of a routine tune-up. Some shops, in fact, simply change the oil and correct the timing of the engine. However, manufacturers, who know their products best, recommend much more: a complete adjustment of the magneto ignition system to restore it to factory specifications, setting of the engine timing, adjustment of the carburetor, and thorough cleaning and inspection of the whole engine, both inside and out.

A tune-up is always done in a fixed order, regardless of the engine's design or use. For example, you must clean or replace the breaker points and the condenser, which precisely control the ignition system, before you set the engine timing. Likewise, the timing must be correct before you can move on to adjusting the carburetor. The check list in the box, opposite, includes all of the jobs in a complete tune-up; they appear in the order in which you should perform them.

Adjusting or replacing the components of the magneto ignition system—the spark plug, breaker points, condenser, coil and magnets—constitutes the bulk of an engine tune-up. (How these parts work together to generate electricity is explained in the box on page 40.)

Breaker points, the pair of electrical contacts that control the ignition system, are opened and closed by a cam—a bulge on the crankshaft—that pushes against them either directly or by means of an intermediary plunger called a push rod. All sets of points, no matter how they are opened or closed, contain one stationary contact—fastened to the breaker-points housing with an adjusting screw—and one movable contact, connected to the stationary one with a spring. Adjustments are always made on the fixed point; the movable point is pushed as far away as possible by the cam, and the fixed point is moved until the gap between the two matches the specification listed in the service manual.

When adjusting the point gap, be sure that the movable point swings freely and that the points align perfectly when they are closed. If they do not, realign the movable point *(page 45, Step 4)*. Also make sure the contacts are not pitted or coated with oil; although you can file points clean, most manufacturers recommend replacing them—and the condenser—after every 150 hours of use.

Finally, check the breaker-points housing; it must be kept in good condition to prevent dust and grime from interfering with the operation of the points. If you find dirt or oil inside the housing, inspect the rubber seals where the breaker wires and the crankshaft pass through the housing. Also inspect the bushing, or metal sleeve, around the push rod, if there is one on your engine. If any of these parts are cracked or worn, you will have to replace them before going on to set the engine timing.

The timing of an engine is determined by the relative positions of the piston and crankshaft, the breaker points, the flywheel magnets and the ignition coil. To regulate timing, you must slightly shift the position of either the plate to which the breaker points are attached or the ignition coil.

For maximum combustion and power, an engine should be timed so that the points open and the spark plug fires just before the piston reaches the top of its stroke—the position called top dead center. The correct timing position is frequently noted by the engine manufacturer with alignment marks on the flywheel and the chassis of the engine. If your engine has no timing marks, you must remove the cylinder head and measure the piston's position with a machinist's rule to put the piston in position for correcting the timing *(page 49)*. Then, to avoid having to open the cylinder head each time you set the timing, scribe your own timing lines with an awl.

Once the timing is set, reassemble the engine and check every nut and bolt for tightness. In most cases, it is sufficient just to make sure they are not coming loose. However, the bolts that hold the cylinder head—and the retaining nut on the flywheel, if the engine has one—are more critical: They must be tightened with a torque wrench exactly to the manufacturer's specification.

Finally, give the engine a thorough cleaning, especially around the cylinder cooling fins, and adjust the carburetor to ensure that the proper mixture of fuel and air is reaching the cylinder. As a part of every second tune-up, disassemble the carburetor and clean it *(page 58)* before adjusting it.

A Tune-up Check List

Below are the 14 steps of a thorough engine tune-up. The instructions for some of the jobs, which are most often associated with the care or maintenance of a specific part, can be found in sections of the book that deal specifically with the part in question; they are indicated by page references. Jobs specifically linked to tune-ups, for which no page reference is given, are explained on pages 40-53.

1 Remove the engine shroud and inspect the starter assembly *(page 84)*.

2 Check for spark *(page 33)*.

3 Test cylinder compression *(page 37)*.

4 Measure the spark-plug gap; adjust as necessary.

5 Remove the flywheel, and check the crankshaft seal and flywheel key.

6 Remove the cover of the breaker-points housing, check the seal and adjust the breaker points; if necessary, replace the points and the condenser.

7 Adjust the engine timing.

8 Remove the cylinder head and replace the head gasket; clean carbon deposits off the piston and valves. Adjust the valve clearance if necessary.

9 Check the coil and ignition wiring.

10 Remove the air cleaner and clean the filter element *(page 28)*.

11 Examine the carburetor linkage and gaskets; adjust the idle-speed and running-speed mixture settings.

12 Change the oil and check the oil filter *(page 27)*.

13 Remove the muffler and examine it *(page 31)*.

14 Clean the cylinder cooling fins and tighten all nuts and bolts *(page 30)*.

Anatomy of an Ignition System

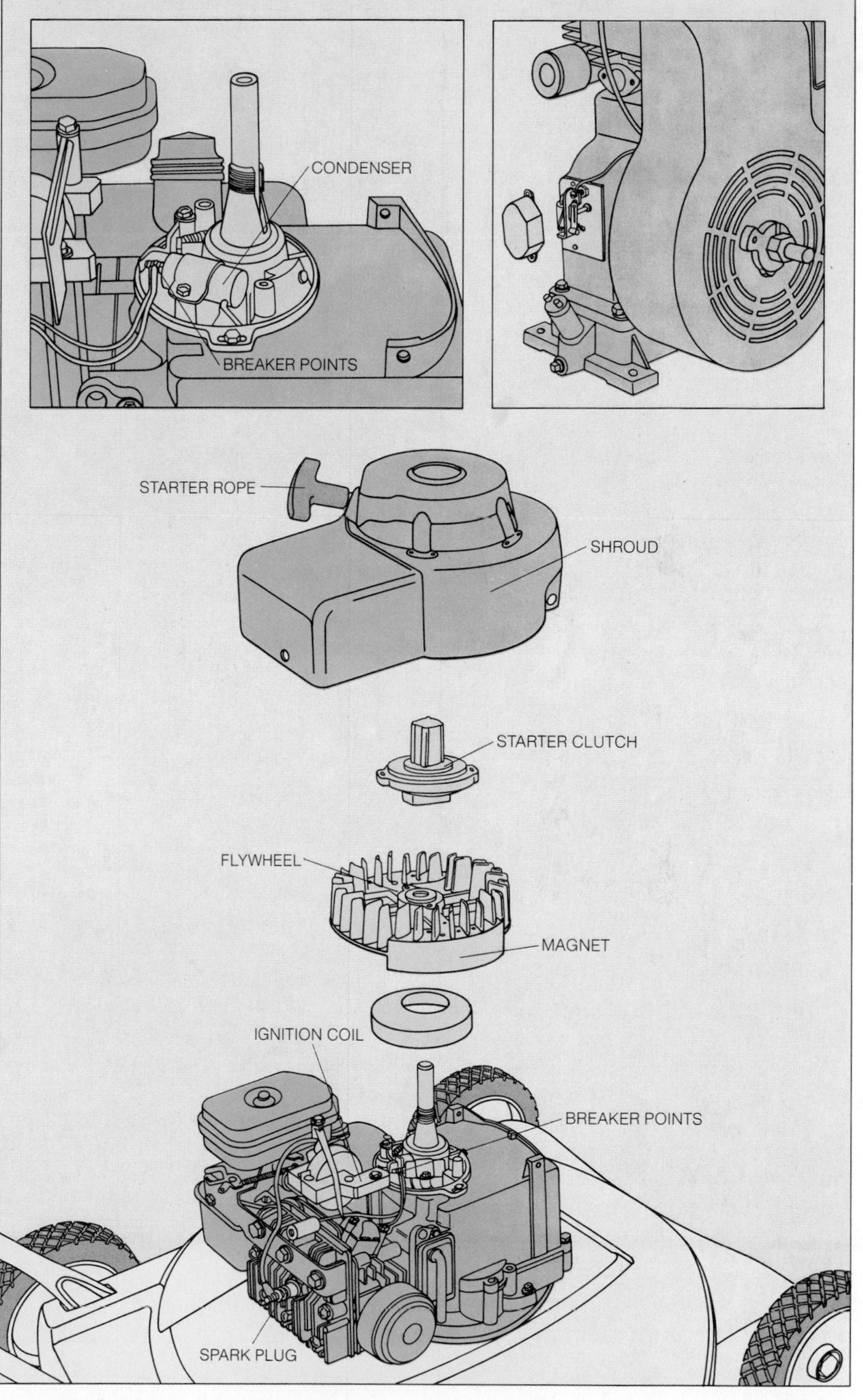

The parts of a magneto ignition system. Under the shroud of a typical small engine lies either a starter clutch, as here, or a ring gear *(page 91)*. These are the parts that spin the flywheel when the starter rope is pulled. Magnets on the rim of the flywheel pass close to the magneto coil as the flywheel spins, generating surges of current to the spark plug *(page 40)*. The breaker points and condenser, next to the coil on this model *(left inset)*, control the timing and intensity of the surges. On other models, the breaker points are located in a box on the chassis *(right inset)* for easier access, instead of under the flywheel.

How a Magneto Generates Electricity

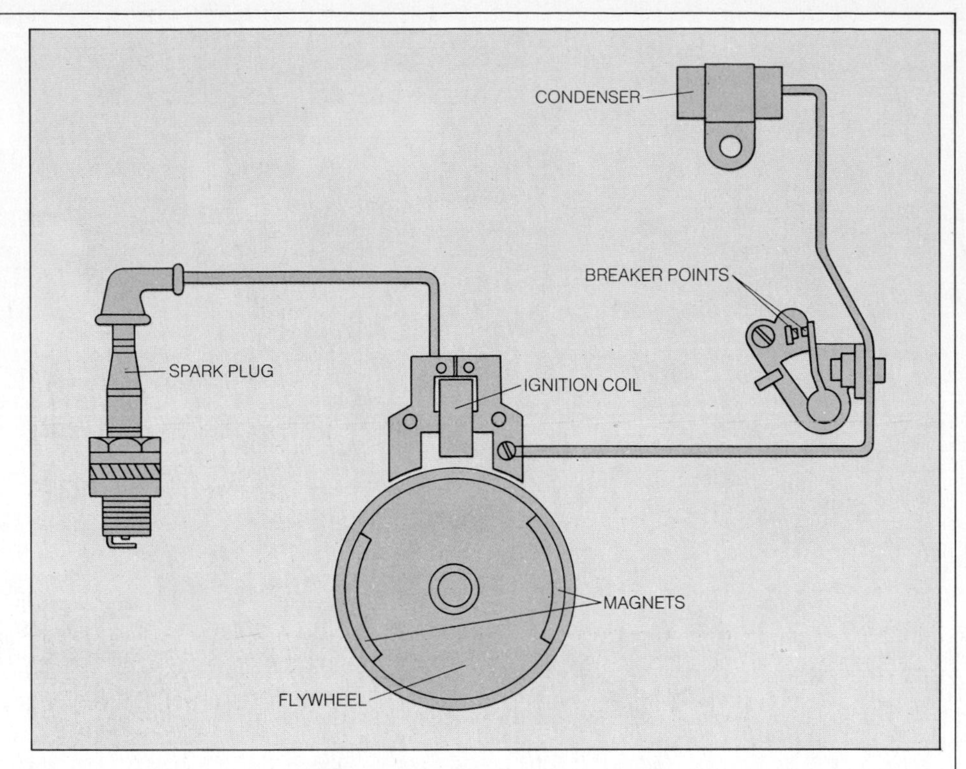

High-voltage surges of electricity—6,000 volts or more—are needed to make a spark plug fire inside a small gasoline engine. These split-second bursts of current are generated—at an astonishing rate of up to 3,000 per minute—by the engine's magneto, which employs a scientific principle called magnetic induction to convert engine motion into electricity.

According to the principle of magnetic induction, the movement of a magnetic field near a conductor, such as copper wire, causes electrons to move along the conductor, establishing an electric current. In a magneto, the driving force comes from a set of magnets attached to the rim of the engine flywheel. As the flywheel revolves, the magnets pass close to an ignition coil, an iron bar wrapped inside two sets of copper wires called the primary and secondary windings.

As the magnets approach the coil, their magnetic field produces a current in the primary winding, causing the winding to act as an electromagnet and envelop the secondary winding in a magnetic field. Abruptly, just as this primary current reaches its maximum strength, it is switched off; the breaker points open, and electrons cannot complete their circuit from the primary winding to ground at the engine block. At the same time, the condenser attracts any last electrons surging toward the opening breaker points, thus protecting the points from being burned by current arcing across them.

The swiftness of the demise of the primary magnetic field is what generates the urge of current in the secondary winding. The strength of these secondary surges is increased further in the hair-thin strands of the secondary winding, which outnumber those of the primary by 100 to 1; the voltage is multiplied proportionately, from 60 volts in the primary winding to 6,000 volts in the secondary. Like a jet of water under pressure, the surging current jumps the gap of the spark plug at the grounded end of the secondary circuit, exploding the air-fuel mixture in the cylinder.

Gapping the Spark Plug

1 Checking the gap. After removing and inspecting the spark plug *(page 34)*, clean any carbon deposits off the electrodes by filing lightly between them with an ignition-points file. Insert the proper wire gauge—usually the .025- or .030-inch one, as specified by the service manual—between the center and side electrodes of the plug. The gauge should drag slightly in the gap between the electrodes. If the gauge passes through the gap without any drag, or if it is difficult or impossible to insert, adjust the size of the gap as demonstrated in Step 2.

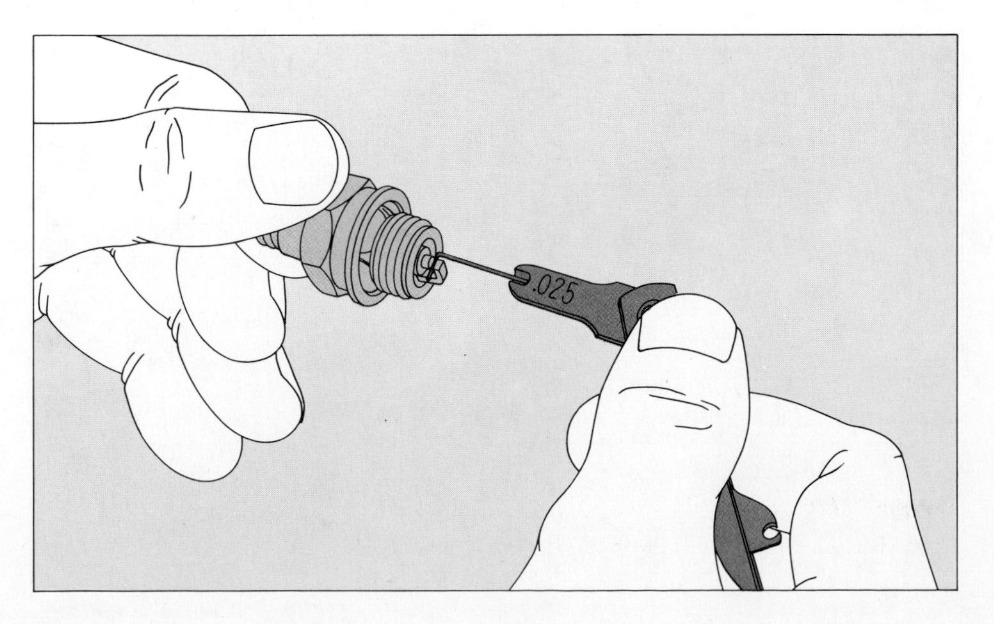

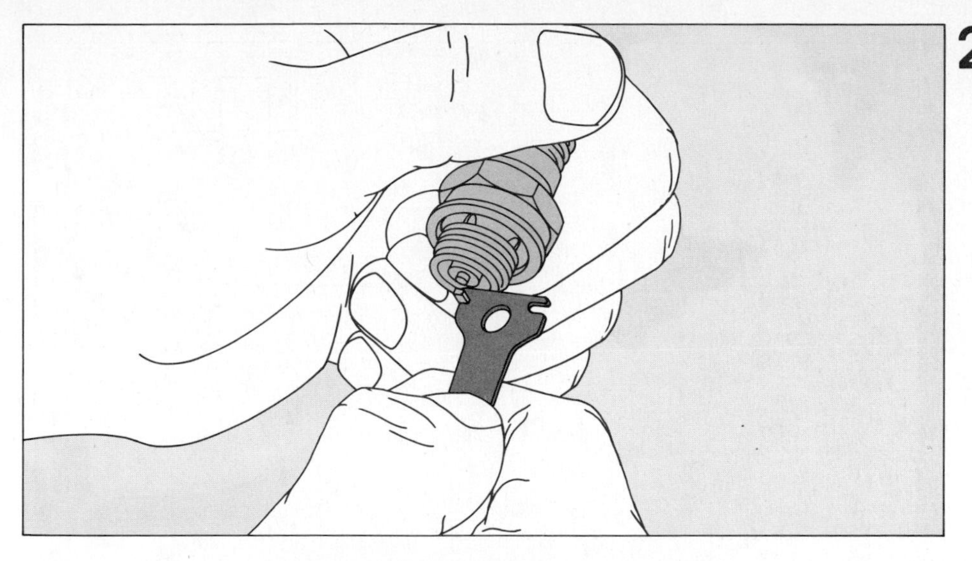

2 **Adjusting the gap.** Slip the slotted bending tool of the wire gauge over the side electrode of the spark plug, and bend the electrode as necessary to widen or narrow the gap. Check frequently with the wire gauge while bending, until the gap is precisely the right size. Take care not to twist the electrode; the faces of the two electrodes should be parallel.

If your wire gauge does not include a bending tool, widen the gap with a small screwdriver; do not pry against the breakable porcelain insulator. If the gap needs to be narrowed, tap the side electrode lightly on a hard surface.

Getting At the Ignition System

1 **Removing the flywheel nut.** After removing the engine shroud to expose the flywheel, unscrew the retaining nut or threaded starter clutch that holds the flywheel onto the crankshaft. To remove a retaining nut *(left)*, insert a wood block between the flywheel and the chassis to prevent counterclockwise rotation of the flywheel. (On some engines, it is also possible to hold the flywheel in position by wrapping a leather belt around the flywheel air vanes.) Once the flywheel is secured, use a socket wrench or an open-end wrench to twist off the retaining nut.

On an engine with a starter clutch *(bottom left)*, secure the flywheel in the same way, then tap the metal flanges at the base of the clutch counterclockwise with a wooden dowel and a hammer. Tap firmly on the side of the flange until the clutch spins off the crankshaft.

2 Removing the flywheel. To protect the topmost threads of the crankshaft, thread a nut onto its end. Insert a large screwdriver under the flywheel, and pry it up while lightly tapping the nut with a ball-peen hammer *(top right)*. Slide the flywheel off the crankshaft, noting how the soft metal key fits into the slot of the crankshaft *(inset)*; you must place the same side of the key in the slot when you reassemble the engine. Inspect both the key and the slots on the crankshaft and flywheel for bending or wear. Check the seal at the base of the crankshaft for leakage. Replace the key or seal if necessary. If the slots are damaged, the crankshaft or flywheel will have to be replaced.

If prying and gentle tapping will not work the flywheel loose, you will have to use a flywheel puller *(bottom right)* designed specifically for your engine model; buy the flywheel puller where parts for the engine are sold. Slide the puller down the crankshaft as far as possible, then secure the self-tapping screws into the flywheel recesses beside the crankshaft. Spin the two lock nuts down the screws until the nuts are tight against the flywheel, then turn both lifting nuts alternately, one revolution at a time, to force the flywheel up the crankshaft.

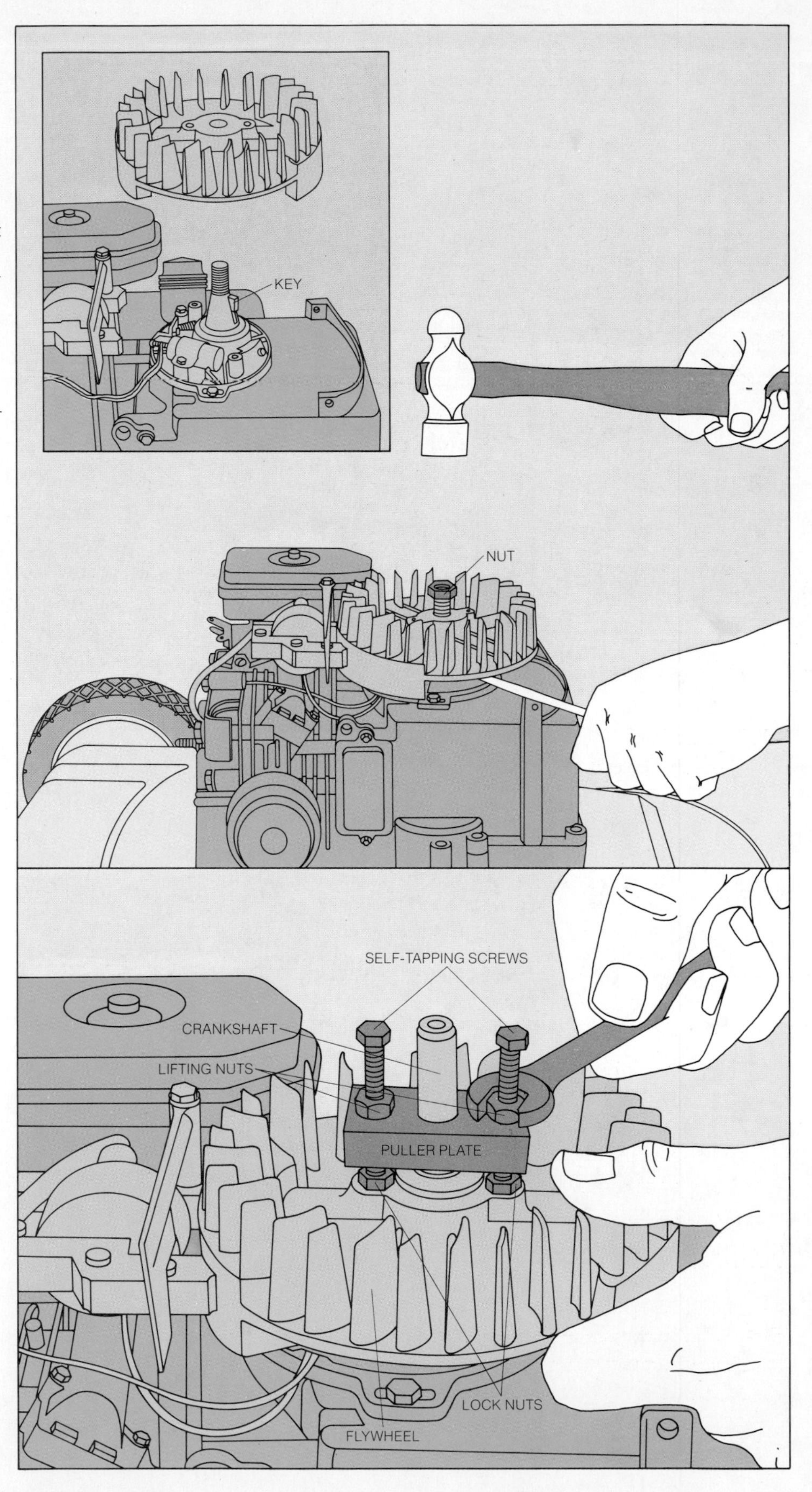

Restoring Breaker Points

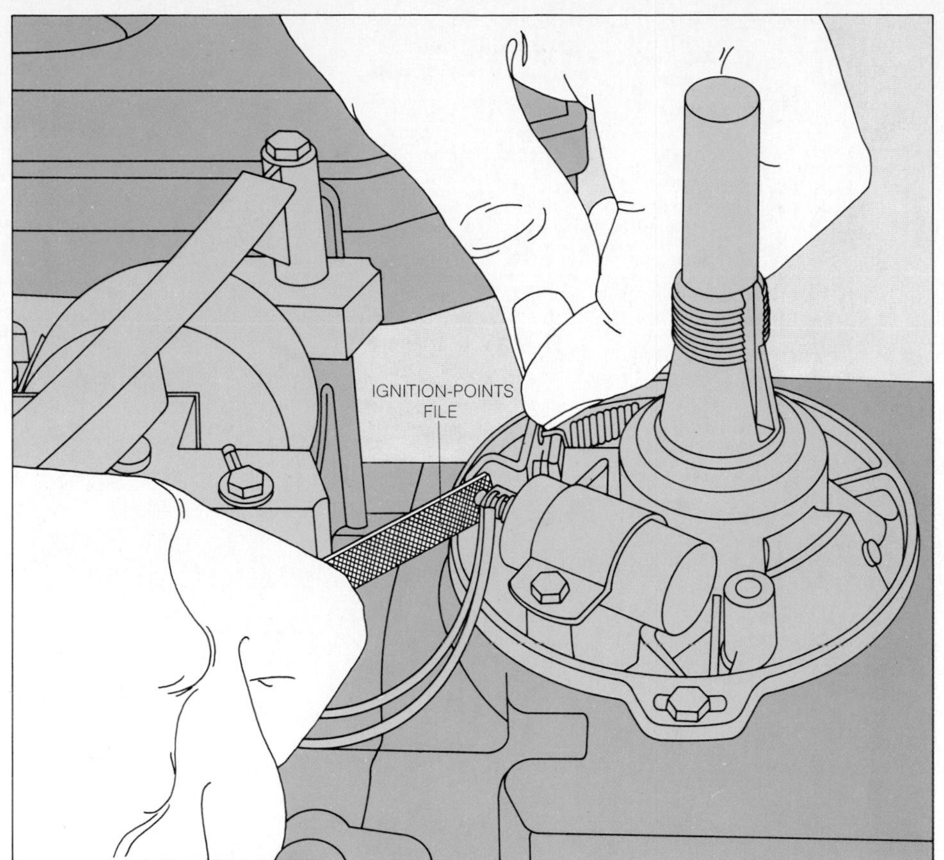

1 Filing the points. Unscrew or unclip the cover of the breaker-points housing and inspect the exposed points. If the points are burned, worn or pitted, remove and replace them and the condenser *(pages 44-45)*. If the points are in good condition, insert an ignition-points file between them; file lightly to remove any deposits. File away as little metal as possible, but make certain that the surfaces of the contacts are completely flat.

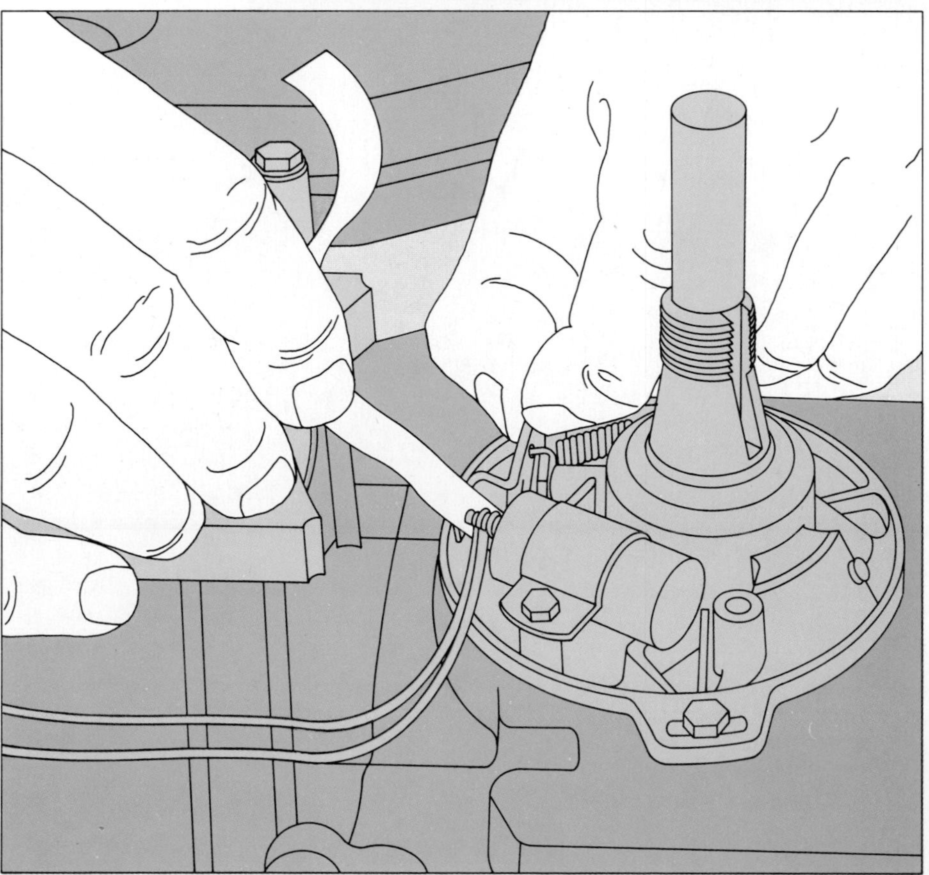

2 Cleaning the points. Soak a strip of bond paper with an electrical-contact cleaner; pull it back and forth between the points, like dental floss, to remove any dust, filings or grease. When removing the paper, open the points with your fingers to prevent the points from tearing off the end of the strip.

Replacing Breaker Points and the Condenser

1 Removing the points. Rotate the crankshaft until the points are opened all the way. Remove the bolt that holds one end of the movable breaker arm *(right)*, and lift the tension spring off its post to free the movable arm. Loosen the condenser clamp and remove the condenser. To detach the wire leads from the condenser, use long-nose pliers to compress the spring that holds the wires, then slip the leads out *(inset)*.

On ignition assemblies with separate condenser and points *(below)*, disconnect the wire leads to the condenser and the coil before removing the points. Take out the screw that holds the points in place *(inset)*, then lift out the terminal to which the wire leads were attached; the points will come out with the terminal.

For an engine with externally mounted breaker assemblies *(bottom right)*, remove the upper mounting screw to free the condenser, then remove the lower mounting screw. Loosen the terminal screw on the points and slip off the leads. Loosen the lock nut on the shaft of the screw that holds the points, then loosen the screw until the bracket that holds the points comes free.

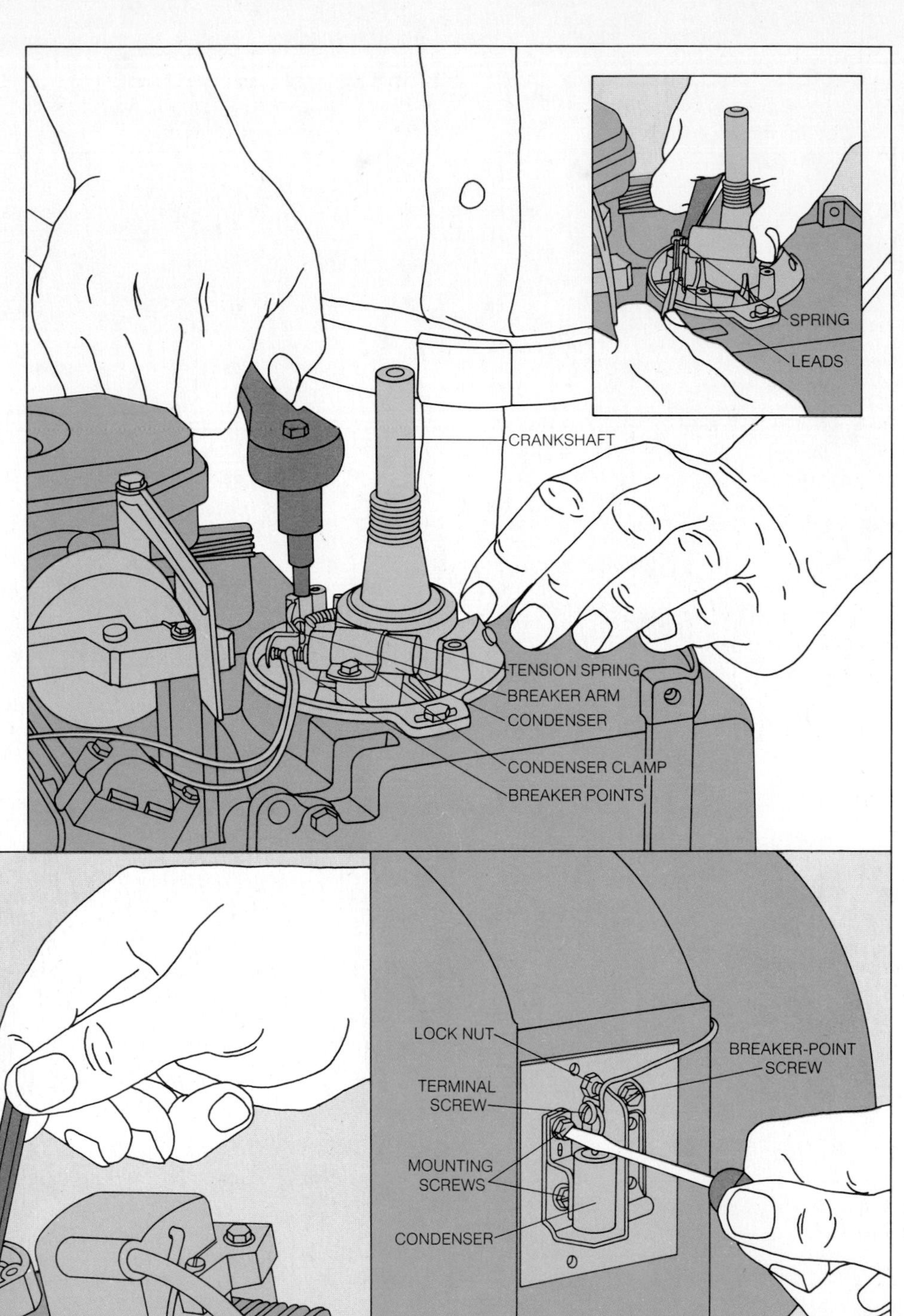

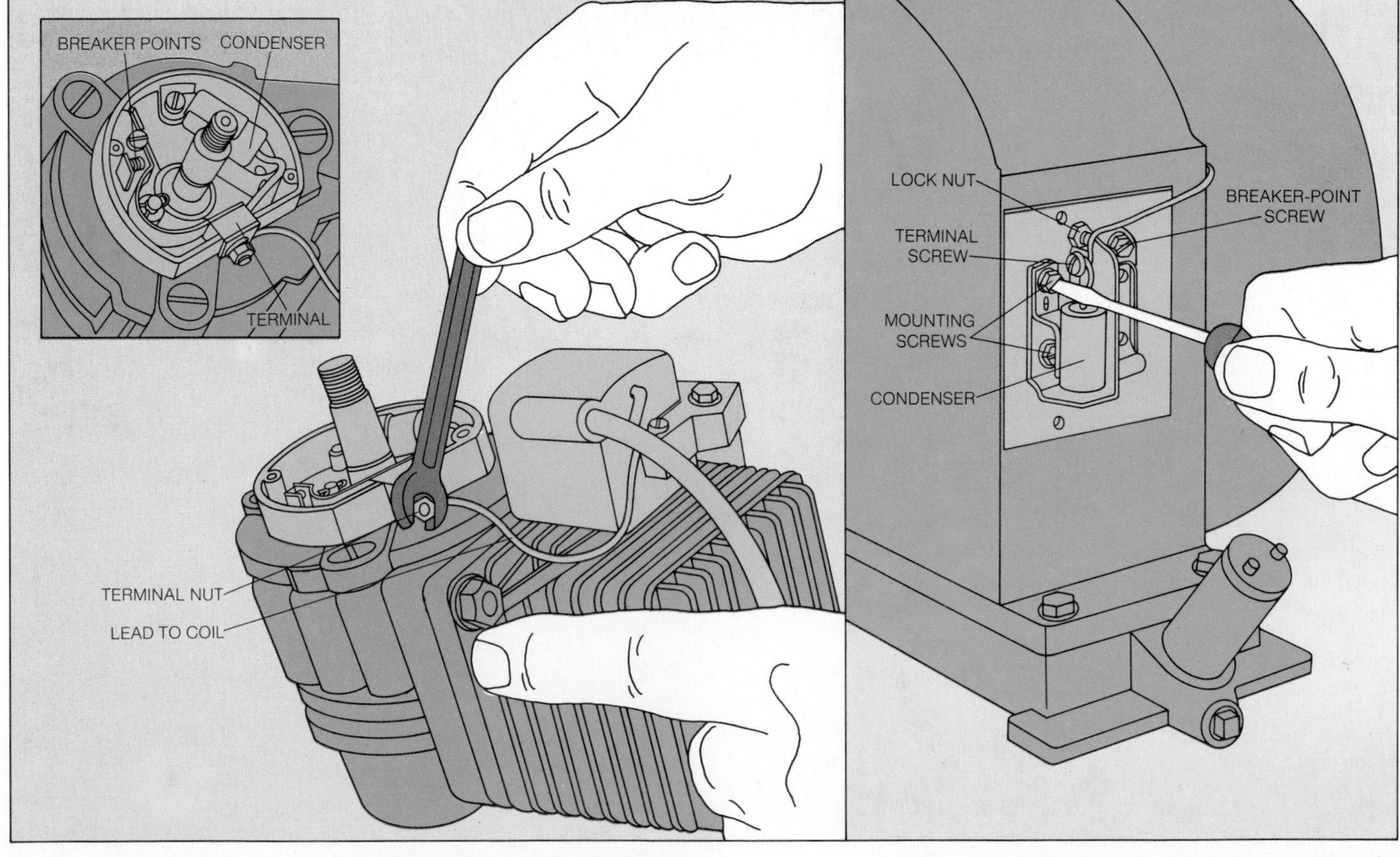

2 Checking the push rod. On ignition systems that use a push rod to move the breaker points, use long-nose pliers to pull the push rod out of its bushing—a sleeve that lines the channel between the crankshaft and the breaker points *(below, left)*. Clean the push rod with fine-grit emery paper, then measure the length of the rod with a micrometer *(below, right)*. If the rod is too short—the proper length is given in the service manual—replace it. Wipe a film of oil onto the rod, and reinsert it so that its grooved end is projecting from the bushing *(inset)*.

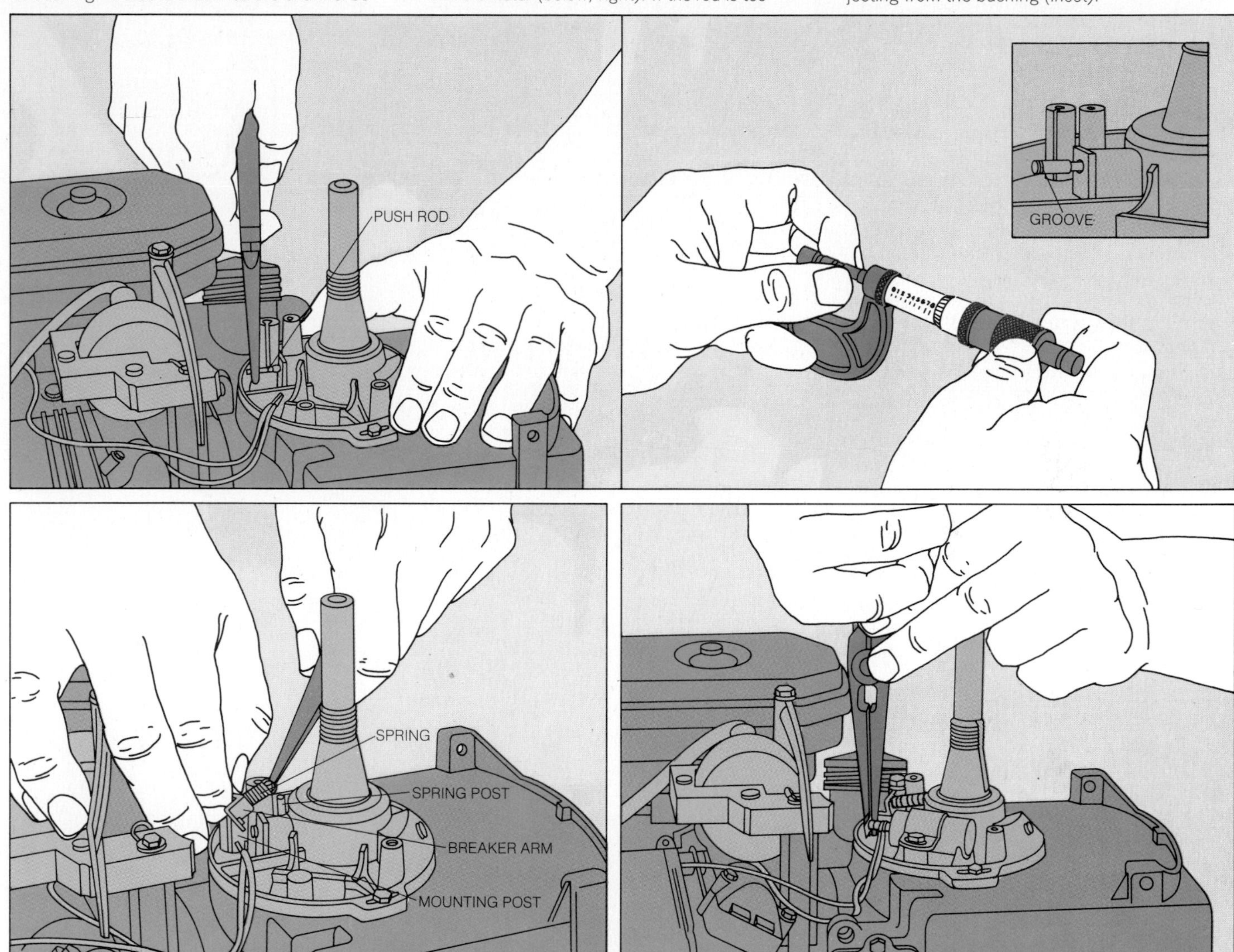

3 Reinstalling the points. Slip the tension spring of the new breaker-points assembly over its mounting post, then insert the end of the breaker arm into the slot on the other mounting post, stretching the spring until the arm is parallel to the floor of the housing and in contact with the push rod. While holding the arm down with one hand, pressing firmly to keep the spring from pulling it out of position, bolt the arm onto the mounting post with your other hand. Insert the leads into the new condenser's terminal. Slide the condenser under the condenser clamp and tighten the clamp screw finger-tight.

On ignition assemblies that have separate condenser and points or have externally mounted breaker assemblies, install the new points and condenser in reverse order from the way in which you removed the old ones.

4 Correcting point alignment. Turn the crankshaft until the points are completely closed. If the contacts are not precisely aligned, use long-nose pliers to bend the arm that holds the movable contact. Do not grasp the contact itself—you could mar the surface. Turn the crankshaft again to reopen the points fully.

5 Gapping the points. Loosen the clamp screw until the condenser can be moved. Insert the proper feeler gauge—.020 inch or as specified by the service manual—between the points, and lever the condenser backward or forward with a screwdriver until you attain the proper gap *(right)*. Tighten the clamp screw.

For systems with separate condenser and points *(bottom left)*, wedge the blade of a large screwdriver in the slot on the fixed point arm. Twist the blade until the points are properly gapped. Carefully tighten the retaining screw without moving the points, then recheck the gap.

For externally mounted ignition assemblies *(bottom right)*, loosen the lock nut on the shaft of the breaker-points adjustment screw, then turn the screw clockwise to increase the gap or counterclockwise to decrease it. When the gap is correct, tighten the lock nut against the crankcase, holding the adjustment screw with the screwdriver so that the gap does not widen.

On any system, rotate the crankshaft a few turns after securing the points, then recheck the gap. If the gap is incorrect, repeat the procedure. If the gap is correct, rotate the crankshaft to close the points, then clean them *(page 43, Step 2)*. Reattach the wire leads from the condenser and the coil. If the leads pass through a plastic grommet, be sure the grommet is properly seated on the housing. If no grommet is present, smear a dab of gasket sealer on the housing where the wires exit. Finally, replace the cover of the breaker-points housing.

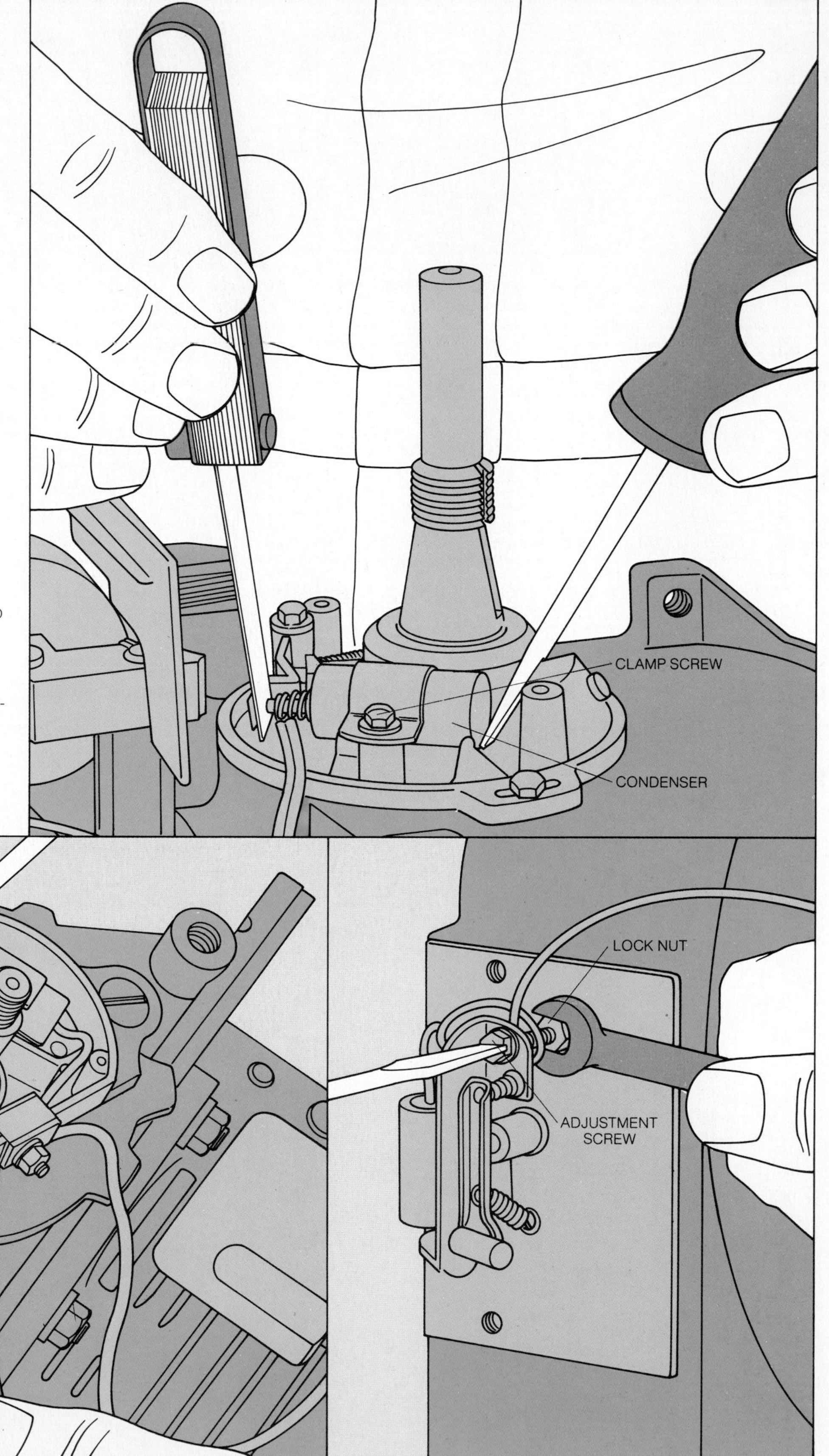

Cleaning the Valves and the Engine Block

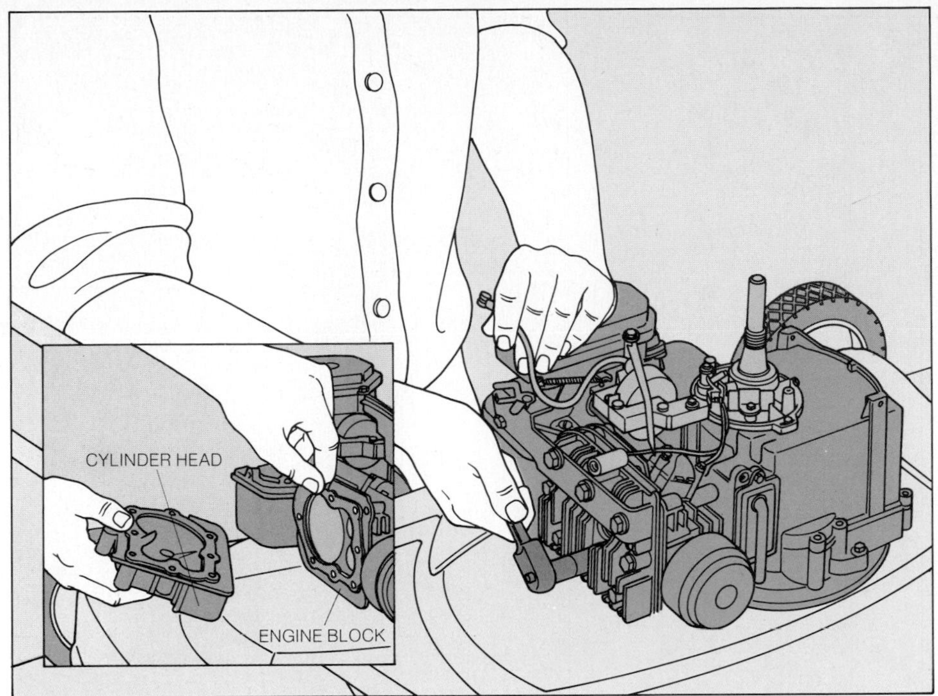

1 **Pulling the head.** Use a socket wrench to remove the bolts that fasten down the cylinder head; to protect the head from stress, which could cause it to warp, loosen bolts in pairs on opposite sides of the cylinder. If the head sticks after you have removed the bolts, tap lightly with a plastic-tipped hammer along the joint between the head and the engine block. Lift off the head and discard the used cylinder-head gasket *(inset)*.

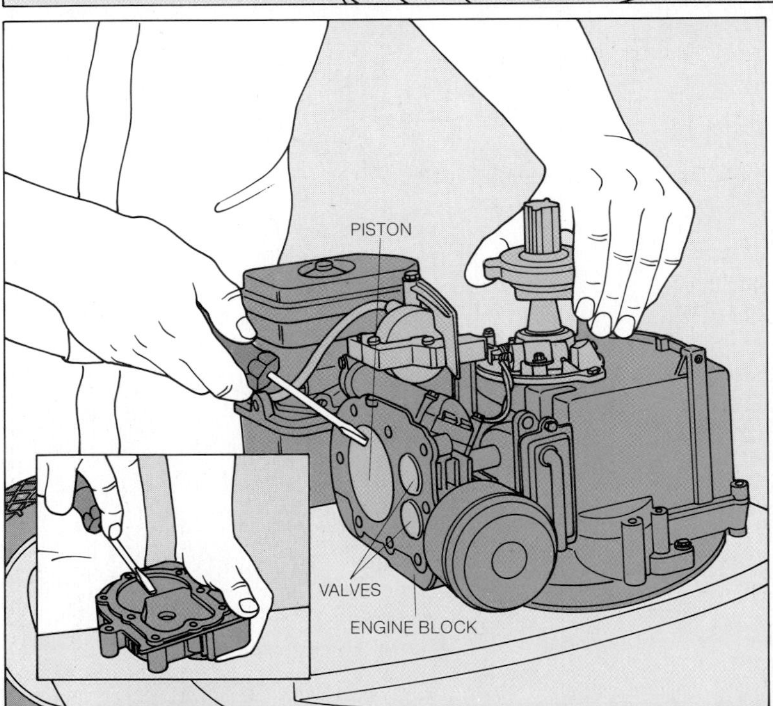

2 **Scraping off the carbon.** Rotate the crankshaft until the intake and exhaust valves are fully closed and the piston is at the top of its stroke—as near to the open end of the cylinder as it travels. Scrape the tops of the valves and the piston with the tip of an old screwdriver to remove any carbon build-up. Do not scrape the surface of the engine block, however; it is more easily scratched than the piston. Scrape the inside of the cylinder head in the same manner *(inset)* but, again to avoid scratches, do not scrape the rim of the head where the gasket fits.

3 **Wire-brushing.** Use a stiff wire brush to remove any carbon from the surface of the engine block and any remaining on the valves, piston and cylinder head. Brush these parts until they gleam; the bristles of the brush, softer than the screwdriver blade, will not cause scratches.

Adjusting the Valves

Correcting the valve gap. Remove the valve-spring cover. Rotate the crankshaft until the valve lifters move away from the valve stems as far as possible. Check the valve gap specified for your engine—your dealer can tell you the proper gap, or you can find it in your owner's or service manual. Insert the appropriate leaf of a feeler gauge between each valve stem and its valve lifter. The feeler gauge should fit snugly without binding. If a gap is too narrow, and if the valves on your engine have adjustment nuts, turn the adjustment nut clockwise to widen the gap. If a gap is too wide, turn the nut counterclockwise; if the nut has already been turned as far as it can go in either direction, the valve will have to be replaced *(page 71, Step 1)*.

If your engine has no valve-adjustment nuts and the valve gap is too narrow, take the engine to a machine shop and have the valve stem ground down .001 to .002 inch at a time until the gap is correct. If the gap is too wide, replace the valve.

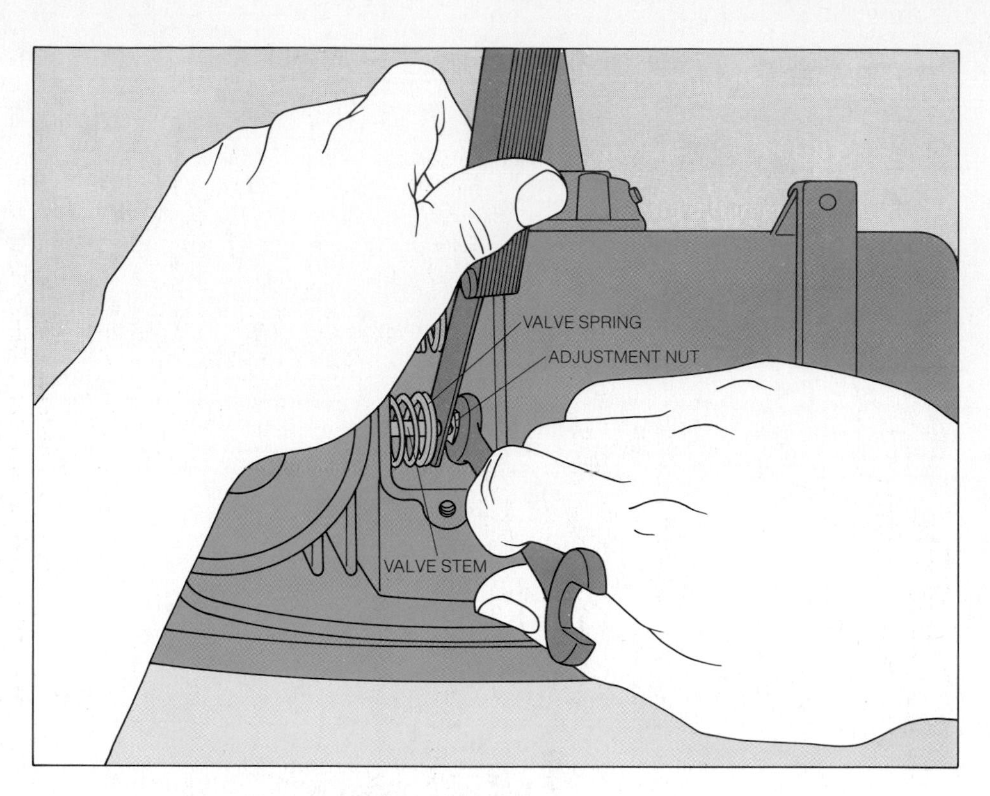

Reinstalling the Cylinder Head

Tightening the head bolts. If your engine has timing marks, replace the cylinder head after adjusting the valve gaps. If your engine lacks timing marks, etch them on the flywheel and engine block, using the techniques shown opposite, before reinstalling the head.

After carefully fitting a new head gasket between the cylinder head and the engine block, put the head bolts back into their holes and tighten them as far as you can with your fingers. Then use a torque wrench to tighten the bolts further, following the tightening sequence recommended by the engine manufacturer. To ensure that the head gasket seals evenly, apply the torque in 5-pound increments, working in sequence on each of the bolts, until you reach the torque specified for your engine.

Etching Timing Marks on the Flywheel and Block

1 Finding the timing position. Rotate the crankshaft until the piston is at the top of its stroke—the position called top dead center. The piston will be either flush with the surface of the engine block or a fraction of an inch below. If it is below, set a straightedge to span the cylinder top, and measure the drop to the piston, using a machinist's rule. Turn the crankshaft in the direction opposite from its normal rotation to lower the piston about 1 inch.

Check with your dealer or consult the service manual to find the piston position for timing; it will be given as a fraction of an inch below top dead center. Hold the straightedge across the cylinder; set the machinist's rule against the straightedge, extending into the cylinder the distance specified for timing. If the piston top was not flush with the surface of the block at top dead center, move the rule farther into the cylinder a distance equal to the measurement of the drop. Holding the rule motionless, slowly rotate the crankshaft in its normal direction until the top of the piston just makes contact with the end of the rule.

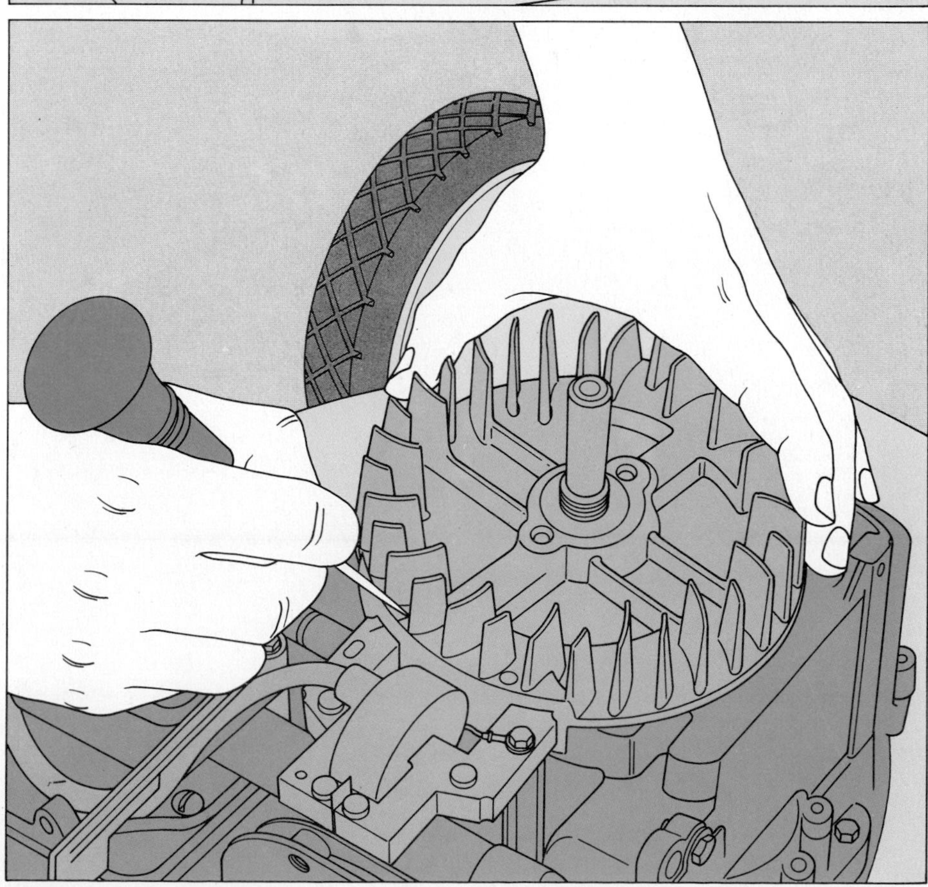

2 Scribing timing marks. Without turning the crankshaft, reinsert the flywheel key just as you found it when you removed the flywheel *(page 42, Step 2)*. Slide the flywheel back onto the crankshaft. With an awl, scratch a line at the edge of the flywheel adjacent to the coil armature *(left)* or adjacent to another convenient engine-block surface, depending on your engine's design. Make a second mark, precisely opposite the first, on the armature or block. Put the cylinder head back in place.

Setting the Timing by Moving the Stator

Using a continuity tester. After aligning the timing marks on the flywheel and the armature or engine block, gently remove the flywheel without turning the crankshaft. Loosen stator bolts *(below, left)*. Clip a continuity tester to the movable arm of the breaker points, and touch the point of the tester to bare metal on the engine block *(below, right)*. Rotate the stator clockwise until the tester light blinks on. Then rotate the stator slowly counterclockwise, stopping just as the test light flickers off. Retighten the stator bolts, taking care not to move the stator.

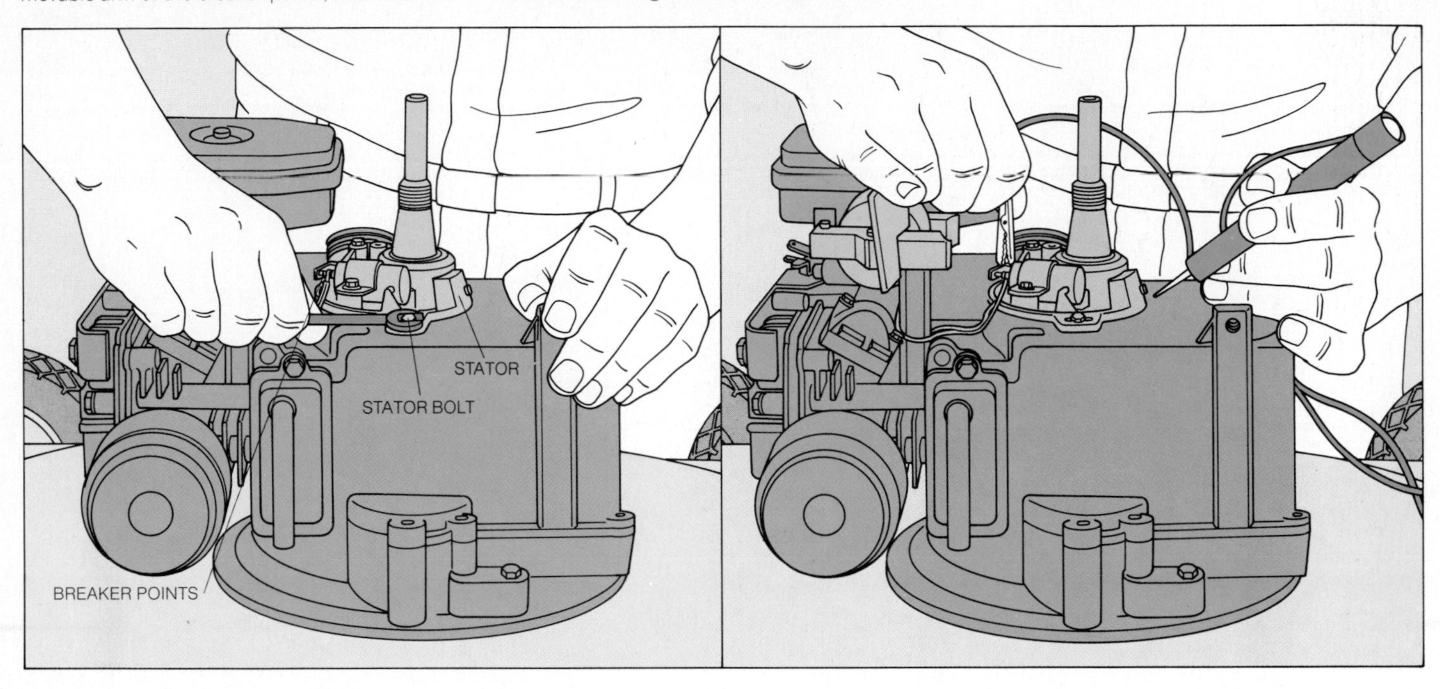

Setting the Timing by Moving the Coil Armature

1 Attaching the continuity tester. Fasten the lead of a continuity tester to the movable arm of the breaker points or to the bare end of one of the wires that are connected to the points. In either case, fasten the lead so that the clip lies flat, leaving room for the flywheel to be installed. Slide the flywheel onto the crankshaft with the flywheel key in position.

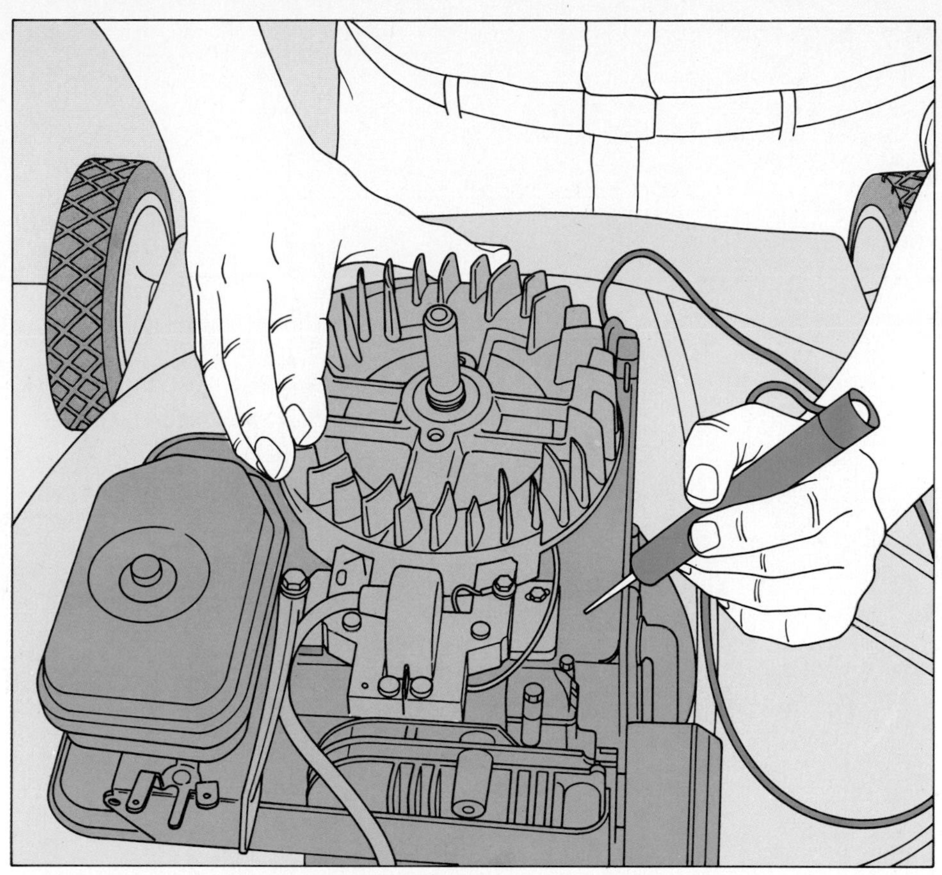

2 Checking the timing. Touch the point of the tester to bare metal on the engine block. If the bulb does not light, turn the flywheel in the direction opposite its normal operating rotation until the light comes on. Then slowly rotate the flywheel in the other direction until the light just goes out. If the timing marks on the flywheel and the coil armature are exactly aligned, the timing is correct. If the marks are not aligned, remove the flywheel, taking care not to turn the crankshaft in the process; disconnect the continuity tester and proceed to Step 3.

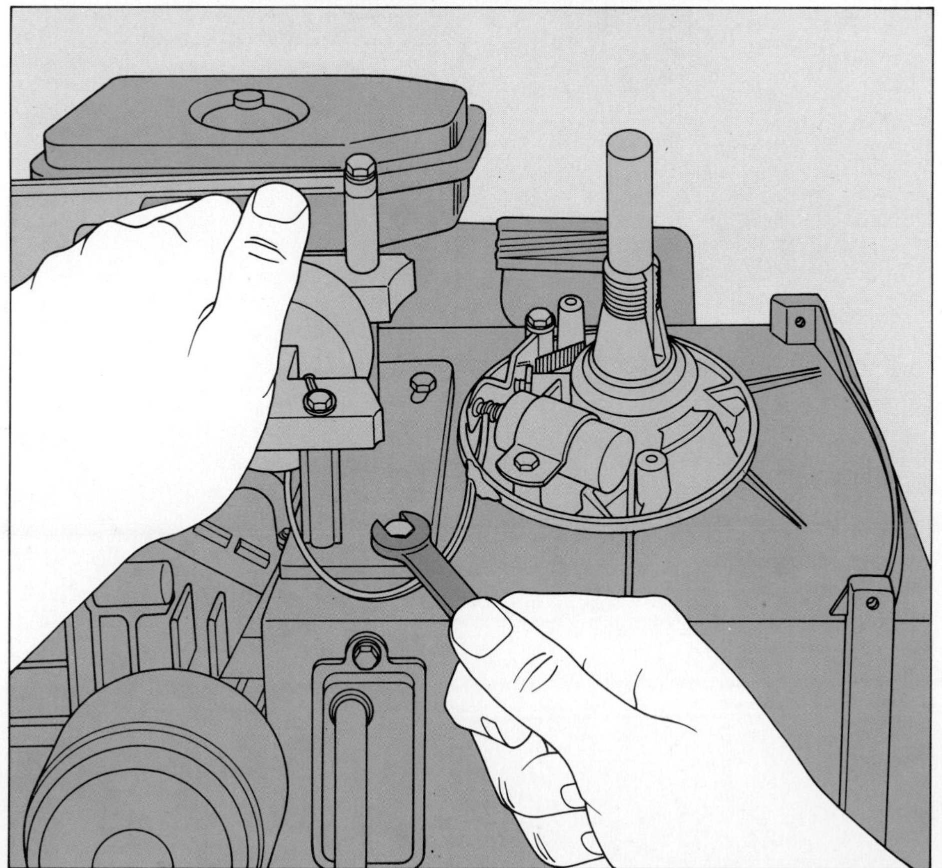

3 Resetting the timing. With an open-end wrench, loosen the bolts that hold the mounting plate of the coil armature to the engine block; loosen the bolts just enough to allow the armature to slide. Reinstall the key and flywheel, again taking care not to turn the crankshaft. Slide the armature to align the timing mark. Once again remove the flywheel, this time taking care not to bump the armature; tighten the bolts that secure the mounting plate.

Reinstalling the Flywheel

1 Testing the magnets. Turn the flywheel upside down so that the two magnets at its base are exposed, then dangle the blade of a screwdriver ¾ inch from each magnet in turn. The pull of the magnets should be strong enough to attract the blade readily; if either magnet fails the test, replace the flywheel.

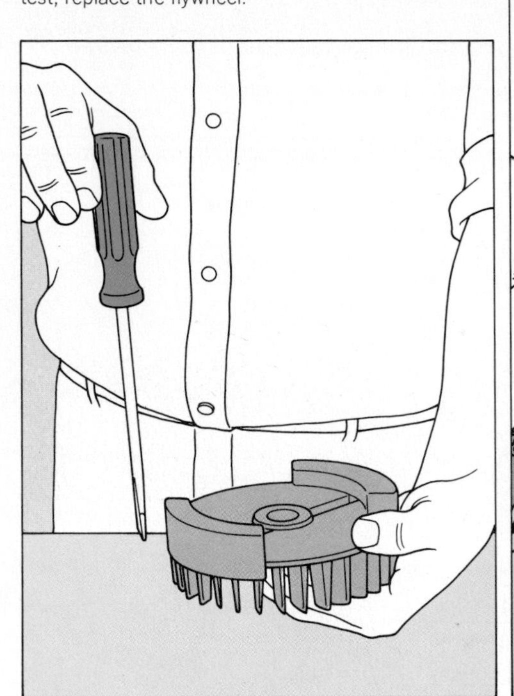

2 Securing the flywheel. On an engine with a flywheel retaining nut, reseat the flywheel and key as they were in Step 2, page 42, then use a torque wrench to tighten the nut to the proper specifications. Wedge a wood block under the flywheel to keep it from turning.

If the engine you are working on is one that is equipped with a starter clutch, twist the clutch onto the threads of the crankshaft until it is finger-tight. Then give the clutch two firm taps with a wooden dowel and a hammer, as shown on page 41, Step 1.

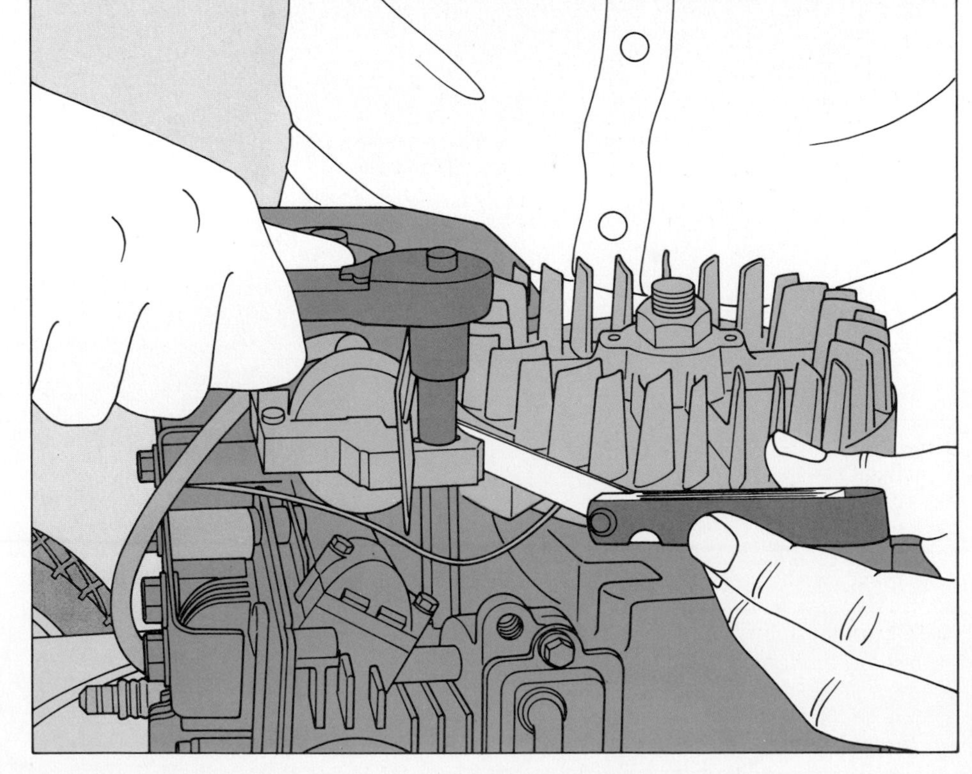

3 Adjusting the armature gap. Rotate the flywheel until one of the magnets is opposite the coil armature. Check the service manual for the proper gap between the armature and the flywheel, and insert the appropriate leaf of a feeler gauge between the magnet and one of the two armature ends. Loosen the pair of bolts on top of the armature and let the magnet pull the armature against the flywheel, sandwiching the feeler gauge between them. Retighten the armature bolt that is closest to the gauge. Withdraw the gauge, and slide it between the magnet and the other end of the armature; then tighten the second armature bolt.

Adjusting the Carburetor

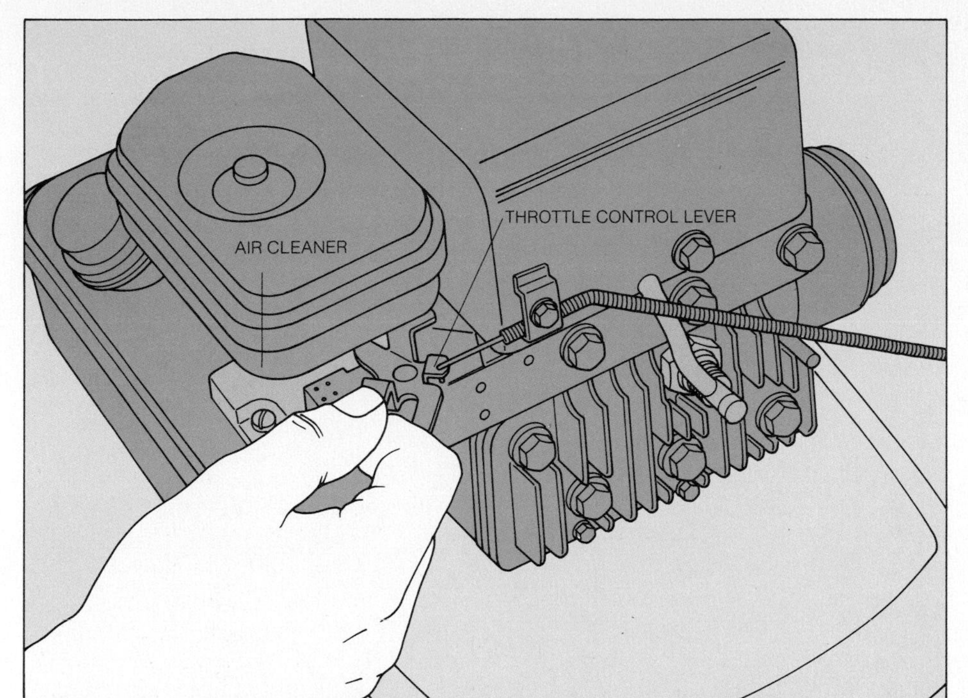

1 Checking the throttle linkage. With the throttle control handle at its lowest setting—OFF or SLOW on most tools—push the throttle control lever back and forth at the carburetor. If the lever sticks, clean it, oil it and work it back and forth until it moves freely and the tension spring can smoothly snap it back to its closed position. Inspect the tension spring—you may have to remove the air cleaner to do so—and replace it if it is stretched or distorted.

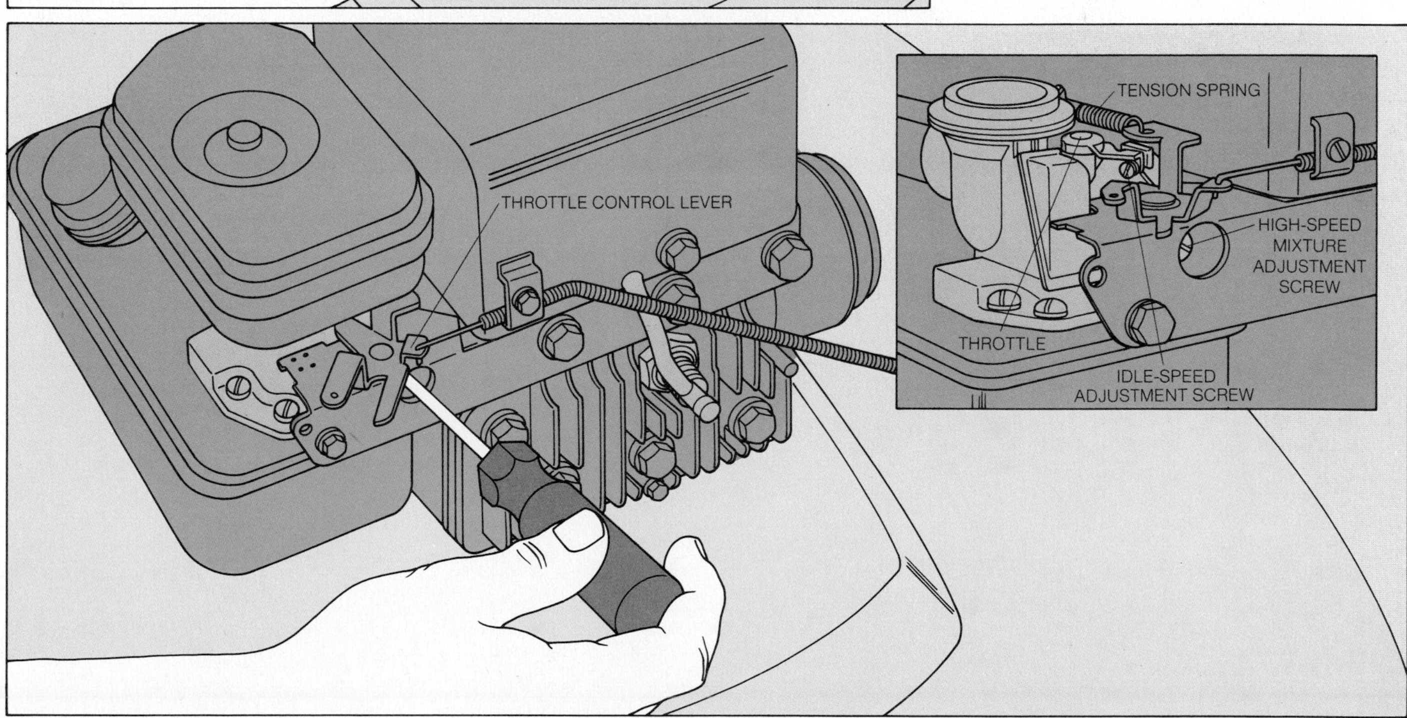

2 Setting the idle and mixture. Check your owner's or service manual for the location of the high-speed mixture adjustment screw. (On some engines it may be called the needle-valve adjustment screw.) Turn the screw clockwise until it just closes; avoid forcing it, however, lest you cause damage to the valve. Turn the screw counterclockwise one and one half rotations, and start the engine.

When the engine has had time to warm up for a couple of minutes, locate the idle-speed adjustment screw (*inset*), and turn it in a counterclockwise direction until the engine sounds as if it is about to stall. Rotate the screw about half a turn clockwise, so that the engine runs evenly but does not race.

With the engine still running, set the engine throttle control to HIGH to race the engine. Turn the high-speed mixture adjustment screw clockwise until the engine starts to misfire. Immediately turn the screw counterclockwise. As you continue to turn the screw, the engine will begin to slow down. Slowly turn the screw in and out until the engine is running at maximum speed.

On some carburetors there is a low-speed mixture adjustment screw in addition to the high-speed and idle-speed adjustments. If your carburetor has this third adjustment screw, reduce the engine speed to an idle and turn the low-speed mixture screw clockwise until it just closes. Immediately turn it back counterclockwise three-quarters to one full turn, until you reach the point where the engine idles most evenly.

Putting a Tool in Storage

Most small engines and yard tools are put into hibernation for a couple or more months a year. But before they begin their seasonal rest, they should be given maintenance work-overs such as the one shown on pages 26-31—with a few added steps. For example, to prevent rust, tools should be thoroughly cleaned and then lubricated. And because gasoline in time evaporates to leave gummy deposits that will clog carburetors, the fuel systems must be drained. These tasks are well worth the time they take, paying off handsomely by extending the lives of the tools.

In the box at right, all of the tasks involved in preparing a machine for storage are listed in the order in which they should be performed. With the exception of changing the crankcase oil, the order is the same for two-stroke and four-stroke engines. (Two-stroke engines, of course, have no crankcase). To perform the jobs most efficiently, read the list over before you begin, and have the tools and materials you will need ready at hand. Do the work outdoors so that gasoline fumes will pose no hazard.

Also consider in advance the best way to drain your gas tank. Any tank can be drained if the machine is lifted and tilted until gas pours out of the filler hole, and on some engines this is the only technique that will drain the fuel system completely. But a less strenuous and less potentially damaging method is to use a suction bulb of the sort used by cooks for basting roasts and turkeys. Almost all the gas can be sucked out with a few dips of the bulb, and the rest can be removed by running the engine until it stops. Neither of these procedures will be necessary, however, if your engine includes either of the design features shown at right, which make draining the fuel tank easier still.

A thorough storage routine leaves few chores to be done when the engine is brought out again—usually a time when you are anxious to put a tool back to work quickly. The few checks that should be performed to prevent unpleasant surprises are included in a second checklist *(opposite, bottom)*.

A Check List for Laying Up an Engine

When you are preparing a machine for off-season storage, save time and effort by performing these tasks in order:

□ With the ignition set at OFF, remove the spark-plug cable and tape it to the tool body. For a four-stroke engine, drain the crankcase oil *(page 27)*.

□ Drain the fuel system. If the engine has a metal gas tank, spray its interior—and the inside of the gas cap—with a thin film of oil.

□ Drain and clean the sediment bowl, if there is one *(page 26)*. Spray the interior of the bowl with oil.

□ Wash the entire machine thoroughly, using a garden hose at high pressure. Use degreaser on stubborn grease spots. For a mower, prop it to one side *(page 27)* and scrape off crusty grass and dirt. Spray the underside with oil. Right the mower and spray the top coverings.

□ Scrub the cooling fins *(page 30)*. If the engine has a housing, remove it and spray the engine with oil.

□ For a four-stroke engine, add fresh crankcase oil. Clean the air filter as shown on pages 28-29. With the filter off, spray oil into the carburetor throat.

□ Lubricate the cylinder *(opposite)*.

□ Lubricate all drive mechanisms and pulleys according to the instructions in the owner's manual. Take care not to get oil on any rubber wheels or belts. Oil wheel hubs and bearings. Lubricate levers, cables and pivot points. Oil the chain, if there is one, with motorcycle-chain lubricant.

□ If the engine has a battery, remove it, disconnecting the cable from the negative terminal first. Clean the terminals with a wire brush, and top off the battery with distilled water. Charge the battery: You can use a trickle charger, which plugs into a regular house receptacle, or take the battery to a shop that services small batteries. Store the battery on blocks or a workbench in a cool, dry place where there is no danger of freezing. Once a month connect it to a trickle charger, if necessary, or plan to have it charged before reinstalling it.

□ Fill pneumatic tires to full pressure, then take the weight off them by resting the body on blocks.

Draining the Fuel System

1 Emptying the fuel tank. If the engine you are working on has a detachable plastic or metal gasoline tank *(above, left)*, start by unfastening the retaining clips or bolts that anchor it in place. Then detach the fuel line from the tank, catching any fuel that may drip from the line in a small can or a paper cup—do not use a plastic-foam container. Lift off the tank and empty it completely.

If the engine has a fuel shutoff valve that is located between the gas tank and the fuel line *(above, right)*, close the valve and remove the fuel line, catching any fuel drips in a catch pan large enough to hold the contents of the fuel tank. Position the pan beneath the valve, then open the valve and leave it open until gas is no longer draining out of it. Close the valve and replace the fuel line.

2 Draining the carburetor. If the engine has a float-bowl carburetor with a small button-like drain valve on the bottom, as shown here, place a paper cup below it, then push up on the valve to empty the bowl. The gas will flow out around the valve. On a carburetor bowl with an adjustment screw, the button-like drain valve is located off-center and is usually smaller than the adjustment screw.

If the engine has a float-bowl carburetor without a drain valve, you must finish draining the fuel system by running the engine until it stops. If the engine has a suction-feed carburetor located atop the fuel tank, the only way to empty it is to drain the entire fuel system by turning the tool on its side and pouring the gas out the filler hole.

Protecting the Combustion Chamber

Squirting in oil. Using a spark-plug wrench or a socket wrench with a spark-plug socket, remove the spark plug from the cylinder head. Squirt motor oil into the empty spark-plug hole—two or three firm squeezes of an oilcan will do. For a two-stroke engine, use oil of the same weight as that normally contained in the engine's fuel mixture. For a four-stroke engine, use oil of the same weight as that in the crankcase. To disperse the oil over the cylinder walls, pull the starter cord or turn the key of an electric starter to turn over the engine a few times. Reinstall the spark plug.

A Check List for Returning an Engine to Service

Just as there is a logical sequence for putting an engine in storage, so is there one for getting it back in business.

☐ Fill the fuel tank on a four-stroke engine with fresh gasoline. Fill the tank on a two-stroke engine with fresh gas and oil *(page 30)*.

☐ Inspect the fuel system for leaks; gaskets and O-rings can rot during storage.

☐ Move all controls, such as the throttle and the height-adjustment levers, to be sure that they function smoothly.

☐ Remove the spark plug and inspect it; clean or replace it, if necessary *(page 34)*. Pull the starter rope or turn the key of an electric-start engine to turn over the engine about 10 times, dispersing the oil in the cylinder. Reinstall the spark plug.

☐ Check the oil level in a four-stroke engine, and top it off if it is low.

☐ For a battery-powered electric-start engine, check the water level of the battery, then make sure that it is fully charged. Reinstall the battery, fastening the positive cable first.

☐ Reinflate pneumatic tires.

☐ On a lawn mower, check and, if necessary, tighten the bolt or bolts that hold the cutter.

3 The Ultimate Job: An Overhaul

Delicate extraction. The jaws of a single-purpose tool called a valve-spring compressor compress the spring that holds an intake valve tight against the valve seat. Like the smaller exhaust valve in the foreground, the intake valve can then be removed from its chamber for cleaning and measuring; if necessary, it can be reground to a precise fit.

Major repair work on an engine, such as a complete overhaul or the rebuilding of a carburetor, reveals one fact that may not be clear during regular maintenance and tune-up: Inside the engine's dirty, heavy-duty casing lie mirror-smooth surfaces and parts that are measured in thousandths of an inch. The precision required to repair these innards of an engine brings the work of the home mechanic almost into the realm of the watchmaker.

Other differences arise from the scope of the job. An overhaul is not a routine matter. It restores the major parts of the engine, carburetor or starter to factory condition and entails replacing or rebuilding any parts—a crankshaft, camshaft, piston or a complete carburetor—that are broken or worn out. In this era of disposable goods, many owners may be tempted to discard an engine that has such serious troubles, but the procedures demonstrated in this chapter can often rescue it.

An overhaul takes time—at least a couple of days for a beginner, with extended breaks for jobs that must be farmed out to a professional, such as reboring cylinders. Approach the task with a composed, positive attitude, and schedule it during the off season (in winter for a mower, summer for a snowblower) when the professional's prices will be lower and promptness less essential. You will find that, in overhauling an engine, mechanical ability counts for less than patience and orderliness. Much of the work is investigative and calls for good powers of observation—in comparing part sizes, noting how things are connected, and looking for signs as subtle as a slight discoloration in a metal surface.

The best approach is a methodical, even plodding, one. Work slowly and with special care in dismantling an intricate assembly such as a carburetor, which contains very small parts; if you go too fast, a tiny component may drop out before you see where it came from or how it was oriented. As you remove nuts, screws, bolts or other small fasteners, use masking tape to attach each one to the part that it came from. (This precaution is particularly important when one part of the engine contains several fasteners of the same type but of different lengths.) Line up the parts in order as you remove them, perhaps keeping whole groups together in plastic trays or shoe boxes. Then, for reassembly, simply start with the parts at the end of the line and work backward.

Much of the work of an overhaul, from beginning to end, involves cleaning. Start with a clean work surface, clean the outside of the engine before disassembling and clean each part as you work on it. Do not stint this phase of the job. When you finish, your engine will not only perform like new, but look new as well.

Rebuilding the Innards of a Carburetor

Because it supplies the mix of fuel and air that feeds combustion inside the cylinder, the carburetor is the heart of any engine. If neglected, it can cause such problems as erratic performance and loss of power *(chart, page 60)*.

Although a carburetor's inner workings—a cramped maze of passageways, tiny springs, diaphragms and valves—may seem daunting at first, this part of the engine is surprisingly easy to care for. Aside from minor adjustments each time you tune the engine *(page 53)*, a carburetor needs special attention only when it exhibits a serious malfunction.

At that point you will need to disassemble the carburetor, replace some of its more delicate parts and clean those not replaced. If you take apart the carburetor in an organized way, the reassembly will be straightforward. Once the carburetor is back together, you will have to adjust the needle valves that regulate the flow of fuel at various engine speeds. This entire procedure must be performed with care, but it is not time-consuming. After you become familiar with the make-up of your particular model, a carburetor overhaul should involve only about an hour's labor.

A number of carburetor designs are in common use, but most of them are similar to one or another of the three models shown below and opposite. The overhaul procedure is similar for all three types. The real key to doing the overhaul is to buy the rebuilding kit provided by the manufacturer for your specific model.

A rebuilding kit is usually inexpensive and will help you in two different ways: The kit will contain the gaskets and diaphragms that are likely to tear when you disassemble the carburetor. It will also contain such parts as filter screens, needle valves, springs and O-rings, which the manufacturer considers most likely to wear out. Thus, you can use the kit as reference when you evaluate the components for reuse or replacement. Give special scrutiny to those parts whose replacements are contained in the kit.

Cleaning the parts that you do reuse is an important part of the job. Impurities in the fuel may leave deposits that clog the narrow passages of the carburetor. Parts made of metal should be immersed in a solvent and soaked overnight. This will loosen any deposits. To make handling the caustic solvent more convenient, buy an immersion-bath kit from an auto-parts supplier. Such a kit consists of a can of solvent with a wire-mesh basket that fits inside for soaking and draining the parts. The can reseals tightly for safe storage and reuse.

Be mindful, however, that manufacturers are increasingly using plastic parts, which will not stand up to prolonged soaking in a solvent. Spray-on solvents can safely be used on the plastic.

In order to begin an overhaul, you will first have to remove the carburetor from the engine. In doing so, pay attention to the way the carburetor is linked to the engine's governor *(box, page 69)*; its linkages to the carburetor will otherwise be baffling when it comes time to reassemble the engine.

The Three Basic Carburetor Designs

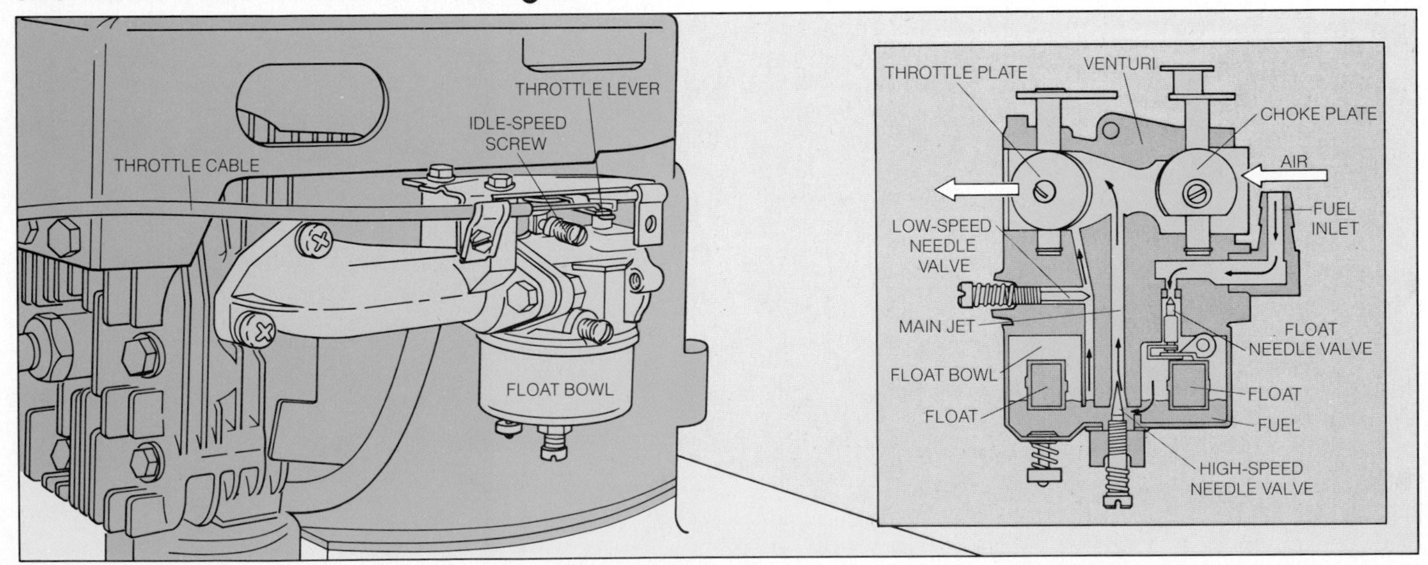

The float-bowl carburetor. Gasoline enters the carburetor at the fuel inlet *(inset)* and drops into the float bowl. The level of fuel in the bowl controls the level of the float which, in turn, pushes the float needle valve up and down to regulate the rate at which fuel enters. The fuel in the bowl then flows through one of two jets to the carburetor venturi, drawn by suction created by the movement of the piston in the cylinder. When the engine is running at high speed, the fuel travels past the high-speed needle valve and up through the main jet into the venturi, where it combines with air rushing past the choke plate through the throat of the carburetor. The air-gasoline mixture is then drawn past the throttle plate and into the cylinder, where it undergoes combustion. When the engine is running at low speed, gasoline is drawn up to the venturi through a different jet, which passes the low-speed needle valve. Both the high-speed and the low-speed needle valves, which perform the function of regulating fuel flow through the jets, can be adjusted during a tune-up by means of screws that protrude from the carburetor. The idle-speed adjustment screw acts as a check on the throttle system to prevent the engine from stalling.

The suction-feed carburetor. The suction-feed carburetor sits atop the fuel tank and draws gasoline directly from the tank rather than from an intermediate container such as a float bowl *(opposite)*. A plastic diaphragm seated between the carburetor and the fuel tank *(inset)* is expanded and contracted by suction created by the movement of the piston in the cylinder; the action of the diaphragm draws gasoline up through the fuel pipe, past the needle valve and into the venturi.

Air moves through the suction-feed carburetor at a right angle, entering the air horn at the top, flowing past the open choke plate on its way to the venturi, and exiting as an air-gasoline mixture past the throttle plate on the side, en route to the cylinder. A single, high-speed needle valve regulates the flow of gasoline to the venturi and is adjusted in the same way as the needle valves on the float-bowl carburetor. The idle-speed screw, located on the throttle lever, serves to adjust the speed of the engine's idle to prevent stalling.

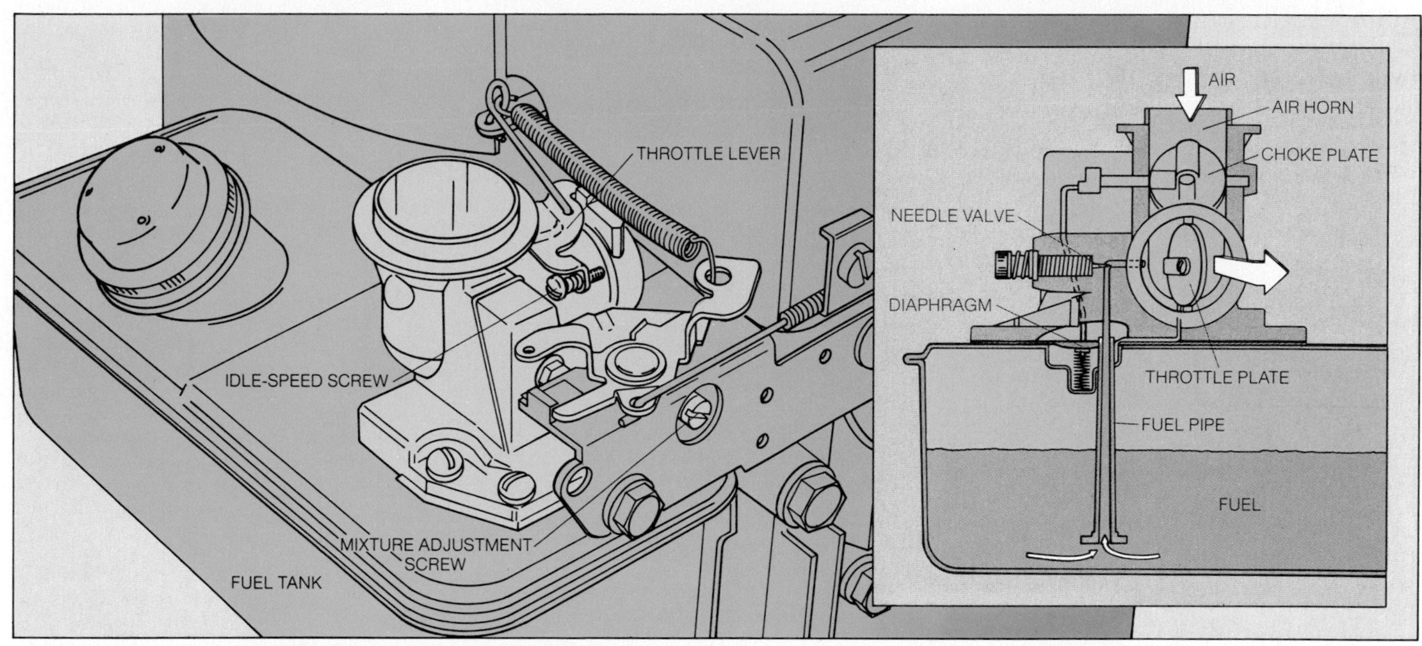

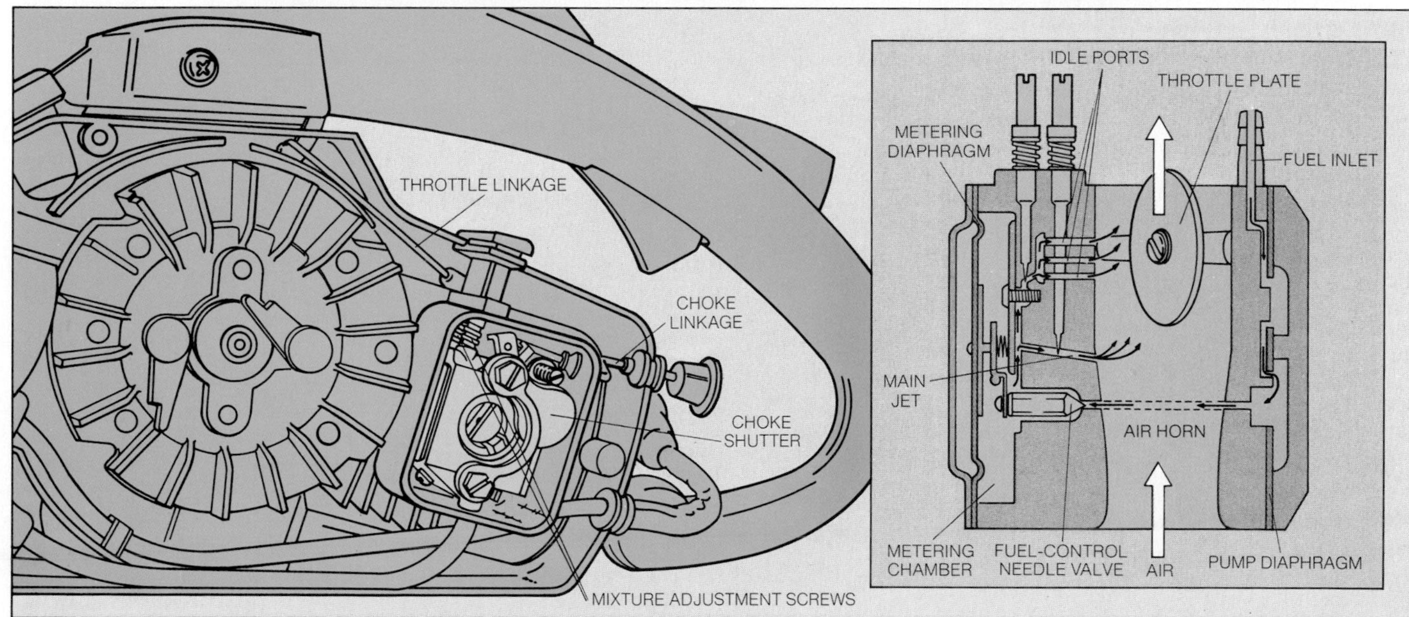

The pressure-feed carburetor. This type of carburetor incorporates two diaphragms: a pump diaphragm and a metering diaphragm. The latter operates a needle valve that regulates the flow of fuel into a metering chamber. The metering diaphragm provides the engine with a steady, regulated fuel supply regardless of the angle at which the engine is held. It is used on the two-stroke engines that provide power for chain saws and weed trimmers, which are operated in a variety of positions.

The inset diagram shows the parts of the carburetor rearranged for clarity. The pump diaphragm, moved by air-pressure changes in the crankcase, draws fuel through the fuel inlet. From there it moves through a passageway *(dotted line)* to the fuel-control needle valve. Low pressure on the cylinder side of the throttle plate raises the metering diaphragm, which opens the fuel-control needle valve and allows fuel into the metering chamber. The angle of the throttle plate, controlled by the throttle linkage, alters the flow of air in the air horn. The choke shutter, controlled by the choke linkage, regulates the amount of air entering the air horn, controlling the richness of the fuel mixture.

Troubleshooting a Carburetor

Symptom	Probable cause	Remedy
Engine idles roughly	Idle speed adjusted incorrectly	Adjust idle-speed screw
	Choke plate stuck	Lubricate choke plate
Engine loses power or stalls when cold	Choke plate stuck in open position	Lubricate choke plate
Engine stalls while operating at low speed	Low-speed mixture screw adjusted incorrectly	Adjust mixture screw
	Float level too high	Adjust float level
Engine loses power when hot	Choke plate stuck in closed position	Lubricate choke plate
	Throttle plate will not open completely	Lubricate throttle plate
	Fuel-tank cap vent clogged	Clean cap vent *(page 35)*
	Diaphragm defective	Replace diaphragm
	Air filter clogged	Clean or replace filter *(page 28)*
Engine uses excessive amount of fuel	Fuel leaking from carburetor	Replace gaskets; tighten housing screws
	Choke plate stuck in partially closed position	Lubricate choke plate
	Air filter clogged	Clean or replace filter *(page 28)*
	Float level too high	Adjust float level
	Float needle valve clogged	Clean needle-valve shaft and bore
Engine loses power or runs unevenly when hot	Choke plate stuck in closed position	Lubricate choke plate
	Fuel-tank cap vent clogged	Clean cap vent *(page 35)*
	Governor malfunctioning	Check linkage; have governor serviced professionally
	High-speed mixture screw adjusted incorrectly	Adjust mixture screw
Engine speeds and slows erratically	Governor malfunctioning	Check linkage; have governor serviced professionally

Using the troubleshooting guide. At the first sign of a problem in the fuel system, check to be sure that there is fuel in the tank and that it is flowing properly *(page 35)*, then refer to the guide above. The left column lists symptoms of fuel-system malfunction; the second column lists carburetor problems that may be causing the malfunction. Directly to the right of each cause is the appropriate remedy. Most of the techniques are illustrated on the following pages; page numbers are given for techniques that are found elsewhere in this volume.

Disassembling and Cleaning a Float-bowl Carburetor

1 Removing the carburetor. Unfasten the air filter from the carburetor, twist the bent end of the throttle-cable link out of its hole in the throttle lever, then remove the two Phillips screws that attach the intake manifold to the crankcase. If the intake manifold seems stuck to the crankcase, lightly tap the joint between the two with a plastic-tipped hammer.

2 Disconnecting the fuel line. Holding the carburetor in one hand, press the ends of the fuel-line spring clamp together with pliers, and slide the clamp along the fuel line away from the carburetor fuel inlet. Release the clamp, grasp the hose end firmly with your fingers and, working it gently from side to side, pull it off the inlet.

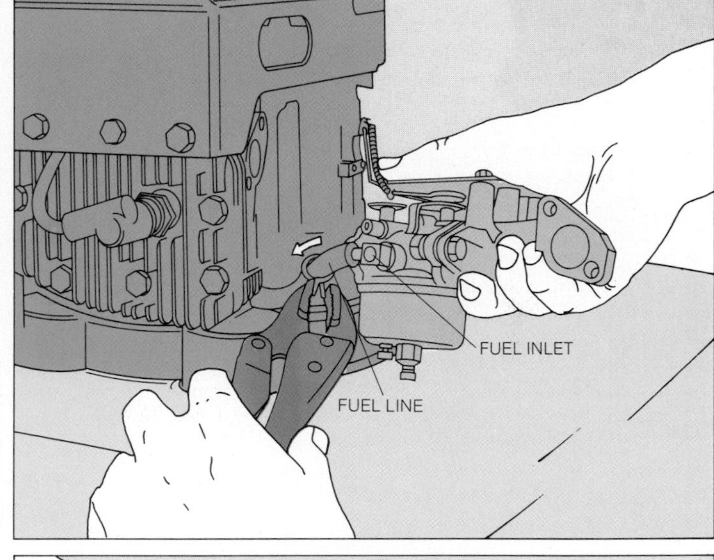

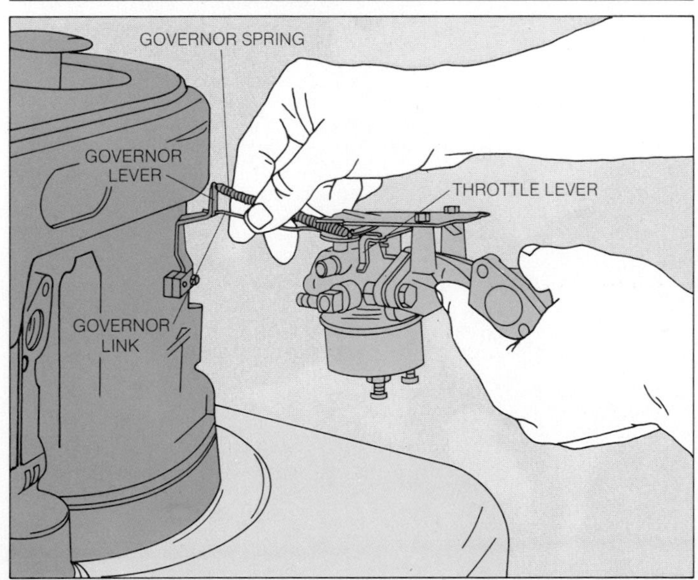

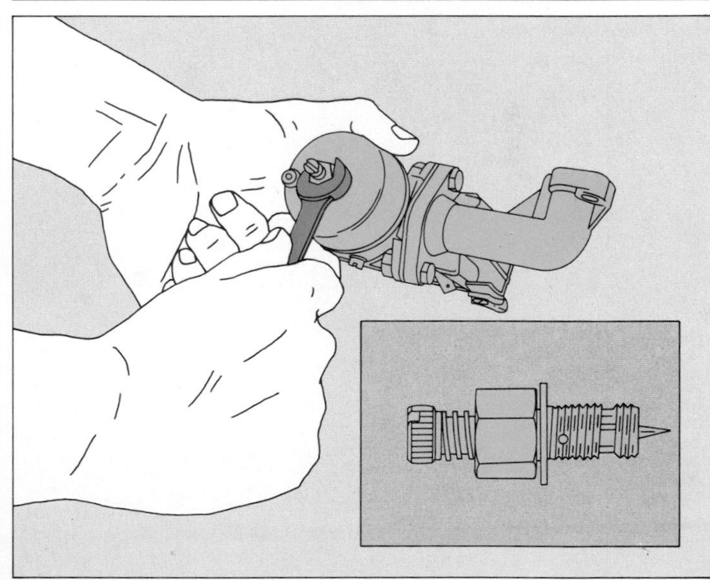

3 Detaching the governor linkage. Maneuver the carburetor so that you can pull the bent end of the governor link through its hole in the governor lever, then twist the governor spring counterclockwise until you can unhook the end from its hole on the same lever; be careful not to stretch or uncoil the spring. Mark the hole for each part with tape to avoid confusion during reassembly. Then pull the opposite ends of the link and the spring from their holes in the throttle lever, and mark these holes as well.

4 Removing the float bowl. Remove the nut from the bolt at the bottom of the float bowl, pull out the bolt and lift the bowl off the carburetor. Examine the small jet hole in the shaft of the bolt *(inset)*; if it is clogged, poke out deposits with the tip of a pin. Remove and discard the gasket that seals the joint between the rim of the bowl and the body of the carburetor; replace it with a matching gasket from the rebuilding kit when you reassemble the carburetor.

5 Removing the needle-valve shaft and float. Pivot the float on its hinge and disengage the valve clip from its tab on the underside of the float with long-nose pliers. Then lift the needle-valve shaft out of its bore *(below, left)*. Set the needle valve aside and take off the float by pulling the pivot pin out of its housing, again using long-nose pliers *(below, right)*.

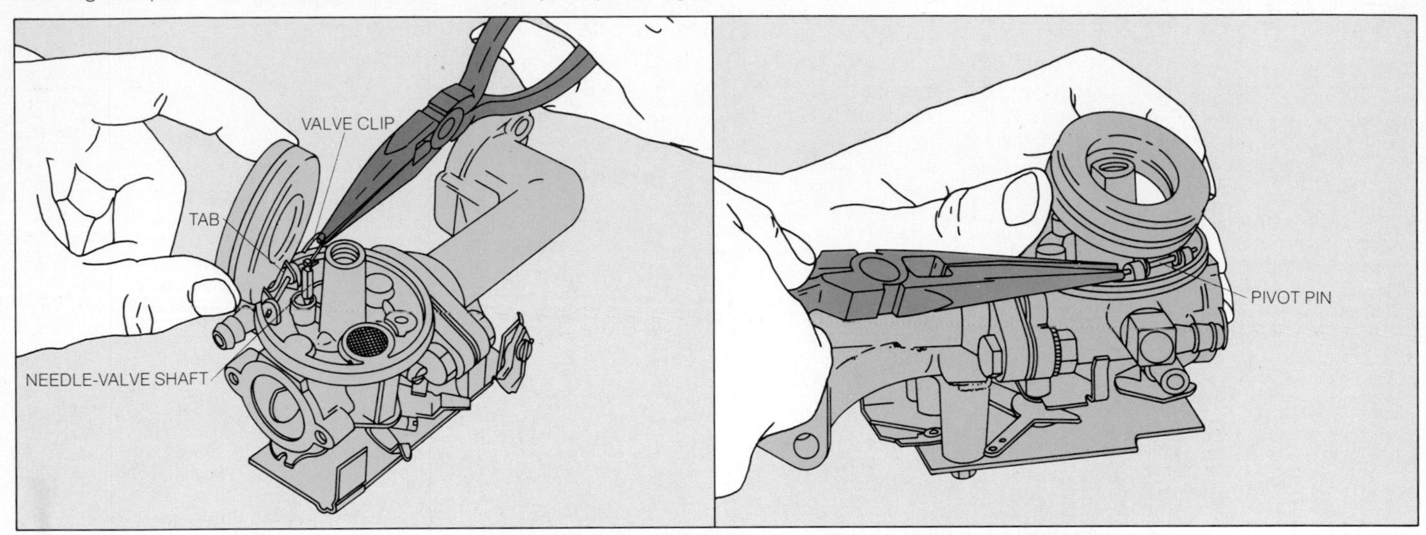

6 Removing the filter screen. Pry the filter screen out of its recess in the carburetor body and replace the screen with the matching one from the carburetor rebuilding kit. Be careful not to press the edges of the new screen down below the recessed ledge in its opening. If the replacement screen is made of nylon, wait to install it until after immersing the carburetor in solvent *(Step 7)*.

Examine the welch plug on the side of the carburetor beside the throttle lever; discoloration indicates fuel leakage. If the area around the plug is discolored, follow the instructions at the bottom of page 63 after you have cleaned and reassembled the carburetor.

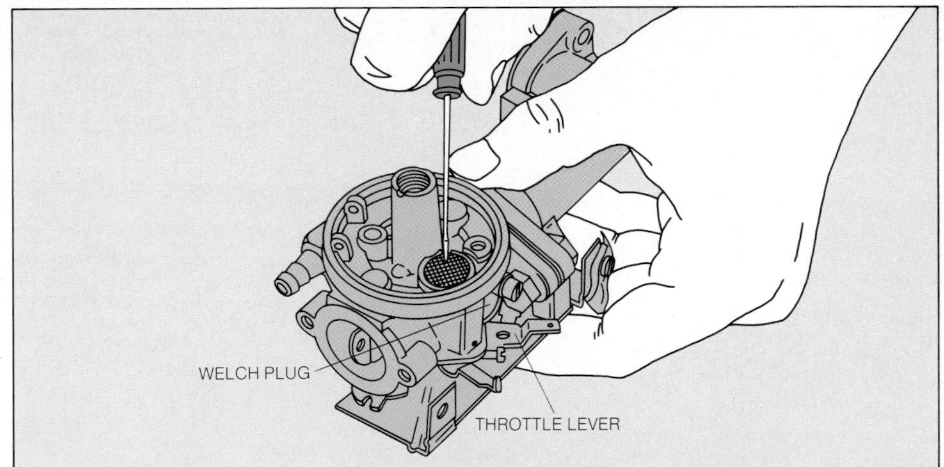

7 Cleaning the carburetor in solvent. Make sure the needle-valve shafts and all of the gaskets, O-rings and other nonmetal parts have been removed from the carburetor and the intake manifold, then set the carburetor parts in the wire basket of an immersion-bath kit and, wearing gloves to protect your hands from the caustic cleaning solvent, submerge the basket in the solvent provided with the kit. Leave the parts to soak overnight. Lift out the basket and swish it a few times in a bucket of soapy water, then in a bucket of clear water. Allow the parts to air-dry; you can dry crevices and interior passages more quickly with a bicycle-pump air hose.

Reassemble the carburetor with all of the new gaskets, screens and O-rings provided in the rebuilding kit. Replace the float in its hinge housing by reinserting the pivot pin; then reseat the needle-valve shaft in its bore, and hook the valve clip back onto its tab on the float.

8 Testing and adjusting the float level. Check the owner's manual for the recommended float height, and slide a drill bit with a matching diameter between the float and the carburetor in a few places *(below, left);* the drill bit should slip in and out, just touching the two surfaces. Make any necessary adjustments by bending the metal tab on the float hinge with long-nose pliers. Reinstall the float bowl. If the owner's manual does not specify a height for the float, measure from the float-bowl gasket to the bottom of the float with a machinist's rule *(below, right)* and adjust the metal tab so that the gap is the same all around the circumference of the float.

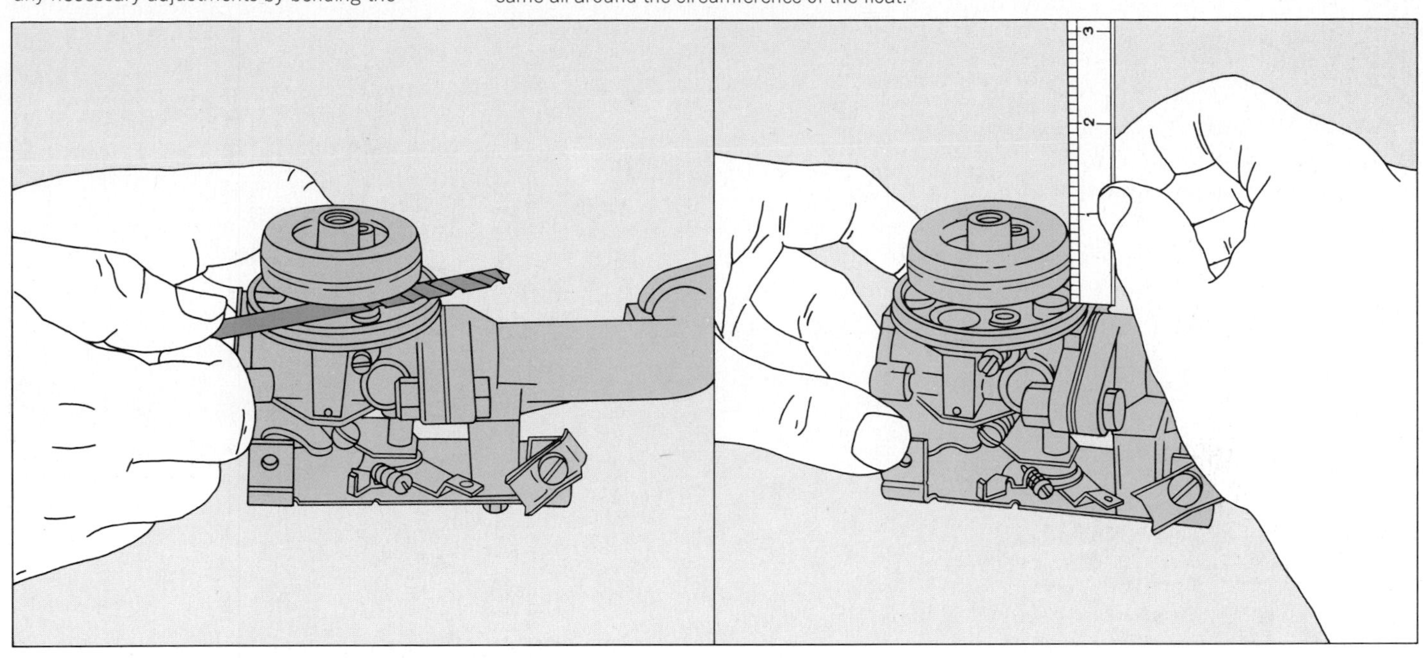

Restoring the Seal of a Welch Plug

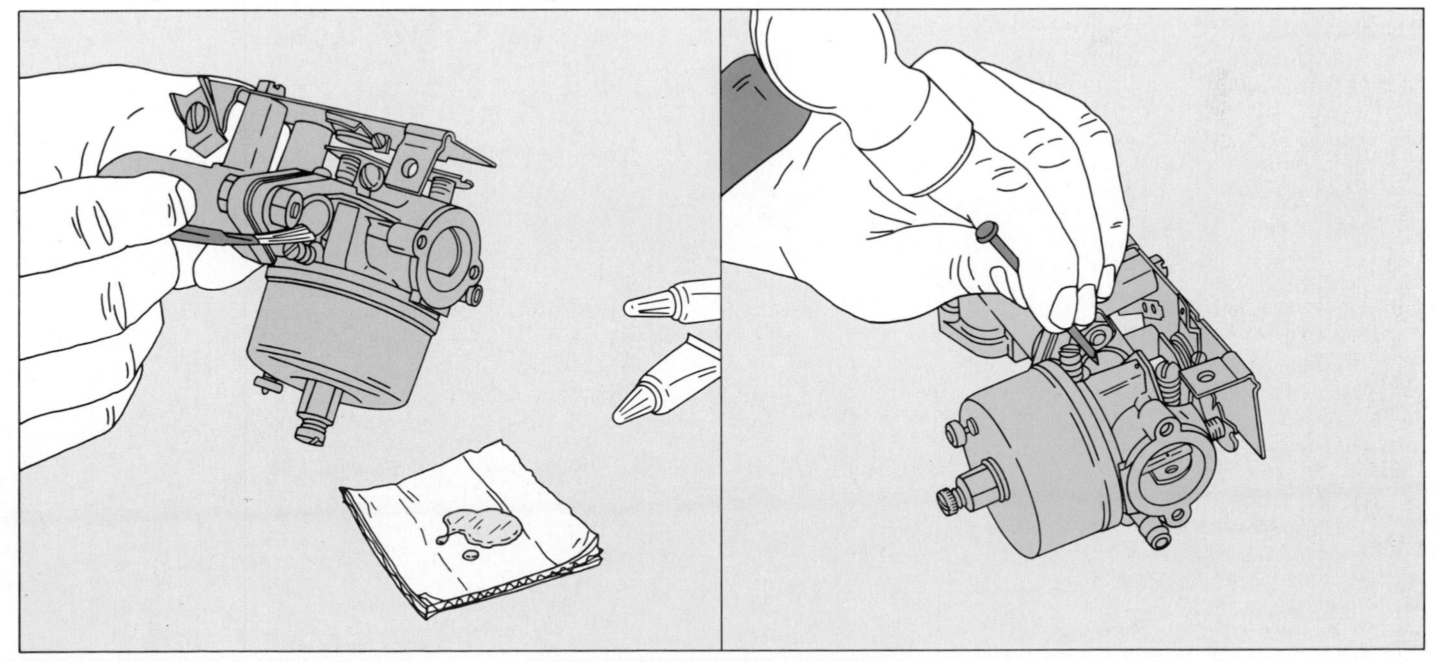

Sealing or replacing a welch plug. If you detected discoloration around the welch plug as you disassembled the carburetor *(opposite, Step 6)*, wait until you have cleaned and reassembled all of the parts, then spread a two-part epoxy cement liberally over the top of the plug and the edge of its bore *(above, left)*. Check the plug after several uses of the tool. If discoloration reappears, gently tap the tip of a nail into the plug *(above, right)*, taking care not to hit the body of the carburetor just below the plug. Then pry the plug out of its bore with the nail. Insert a replacement plug from the rebuilding kit, and tap it into place, flush with the edge of the bore.

Making the Final Adjustments

Checking the throttle and choke plates. With the carburetor detached from the engine, insert your finger into the air horn and push the throttle plate to make sure it pivots easily. If the plate sticks, spray it with a penetrating lubricant. If the plate still moves stiffly, remove its central screw, loosen the lock screw on the shaft of the throttle and remove the plate. Replace it with a new plate. Check and lubricate or replace the choke plate in the same way.

THROTTLE PLATE

Adjusting the carburetor. When the carburetor is remounted on the engine, adjust the idle-speed screw by first turning it counterclockwise until its tip no longer contacts the throttle-lever shaft, then turn it clockwise until the tip just touches the shaft. Finally, turn the screw clockwise one full rotation.

Your float-bowl carburetor will have either one or two mixture screws, depending on the model; the type shown here has two—one for high-speed operation, one for low-speed. Without forcing the screws, turn them both clockwise until they are just tight, then turn each screw one full rotation counterclockwise.

These preliminary adjustments will enable you to start the engine and run it until it is warm; you can then make final adjustments as described on page 53. If the engine will not run without stalling after preliminary adjustments, turn the idle-speed screw one half turn clockwise and try running the engine again.

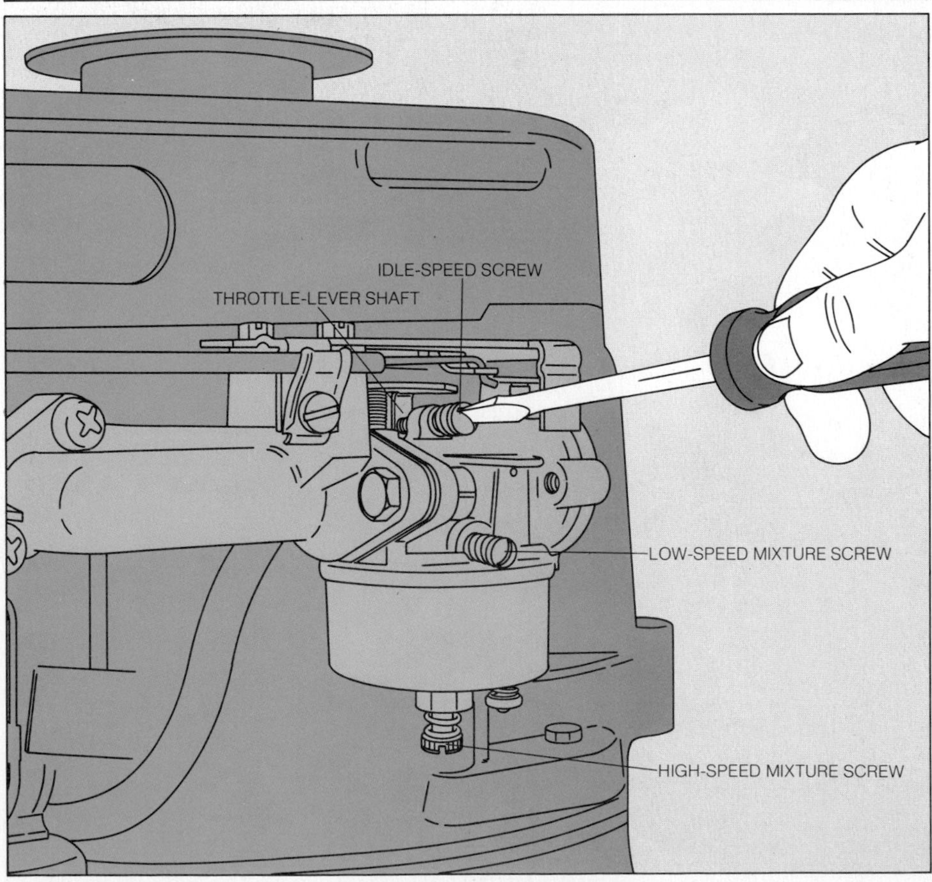

Rebuilding a Suction-feed Carburetor

1 Unfastening the tank and carburetor. Remove the air filter from the engine and disconnect the fuel line at the point where it feeds into the fuel tank; then use a socket wrench to unfasten the bolt that holds the carburetor to the cylinder-head bracket *(below, left)*. Remove the bolt that holds the fuel tank to the crankcase in the same way *(below, right)*. Unhook the end of the governor spring from the control lever, and the governor link from the throttle lever.

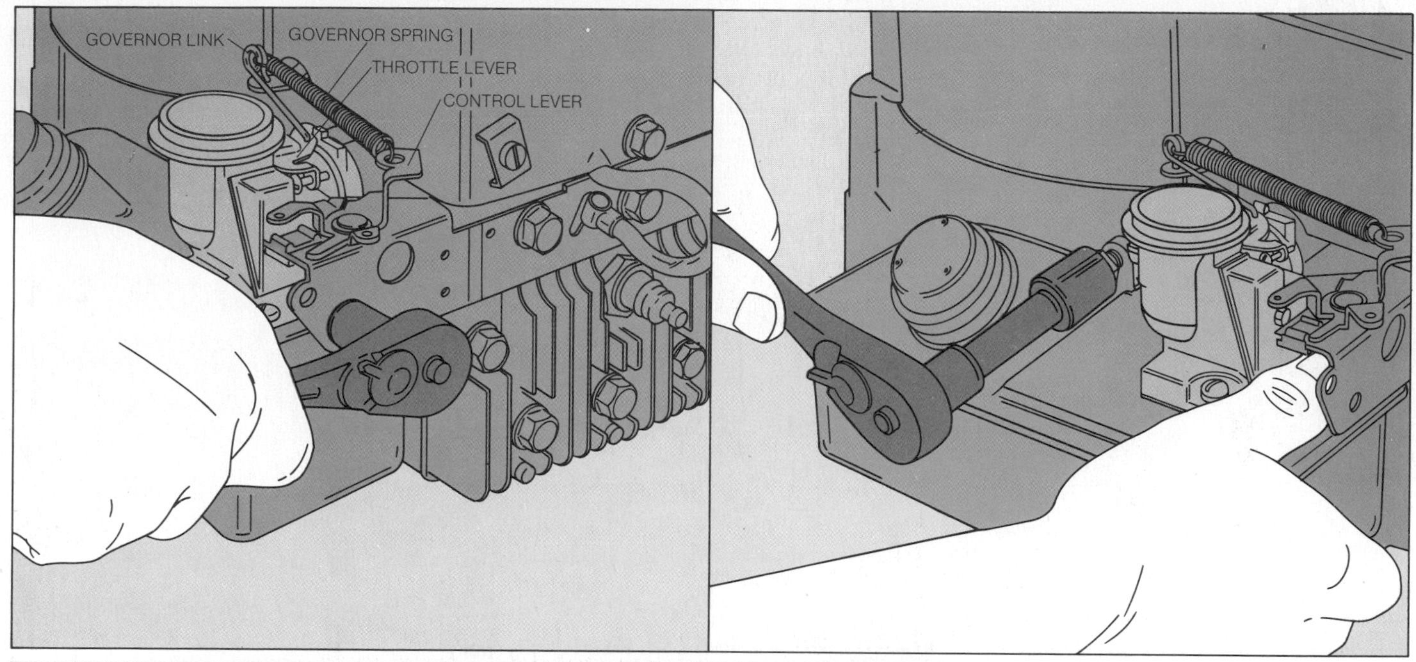

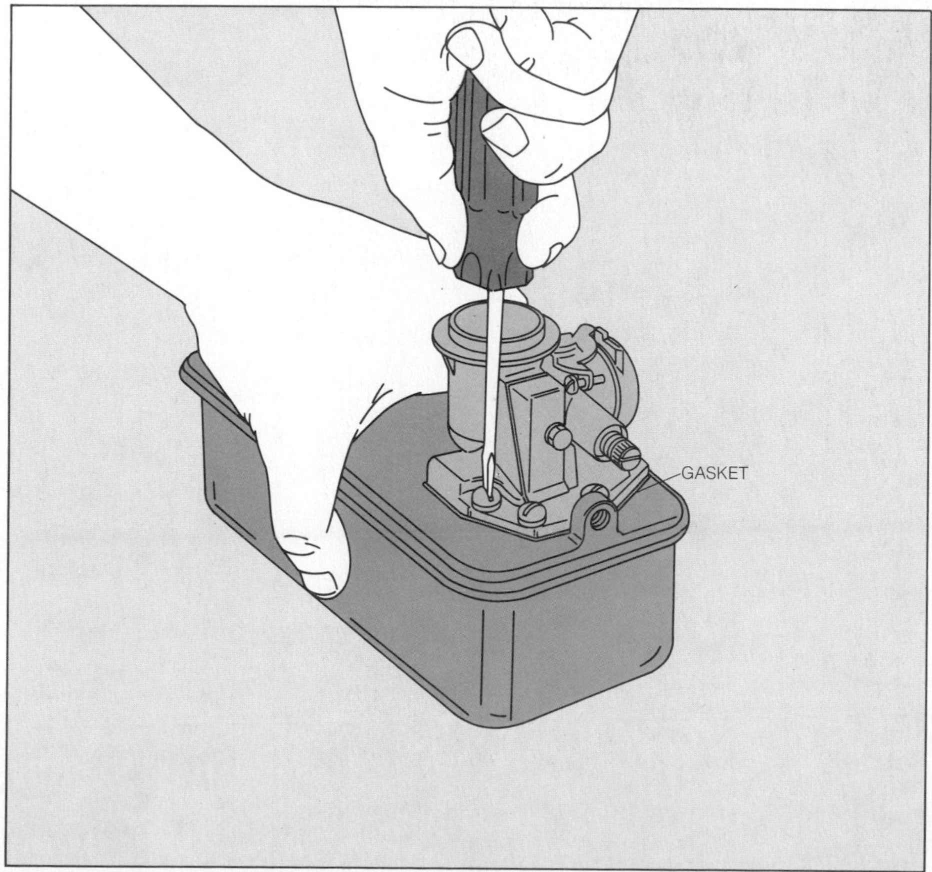

2 Taking the carburetor off the tank. Remove the screws that secure the carburetor to the fuel tank, then carefully lift the carburetor off the fuel tank, peeling the gasket off the bottom of the carburetor as you lift. Replace the gasket with a matching one from the rebuilding kit when you reassemble the carburetor.

3 Detaching the choke linkage. Unscrew the access plate that covers the choke linkage, and pull the end of the choke link out of its bore. Push the link down through the bottom of the carburetor; at the same time slide the diaphragm, which is sandwiched between the choke link and spring, off the fuel pipe. Pull the spring off the flanges on the spring bracket to free the diaphragm. Inspect the diaphragm; if it is torn or if the plastic is at all stiff, replace it when you reassemble the carburetor. Check and service the choke plate as described at the top of page 64.

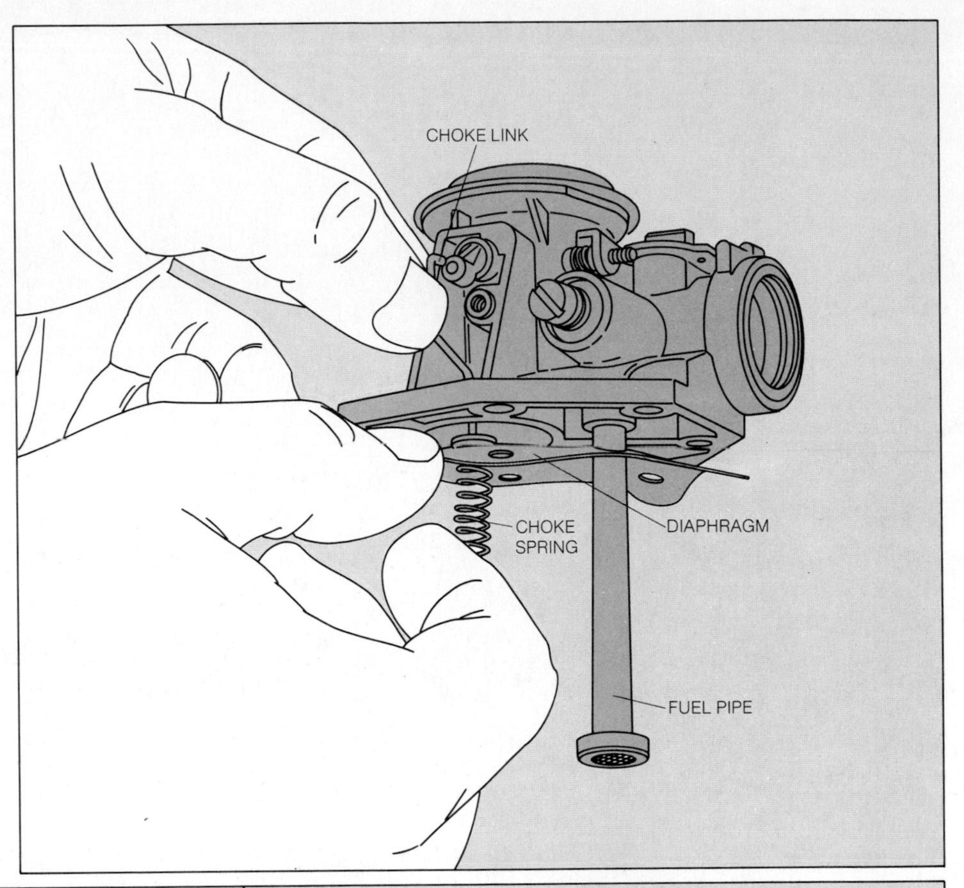

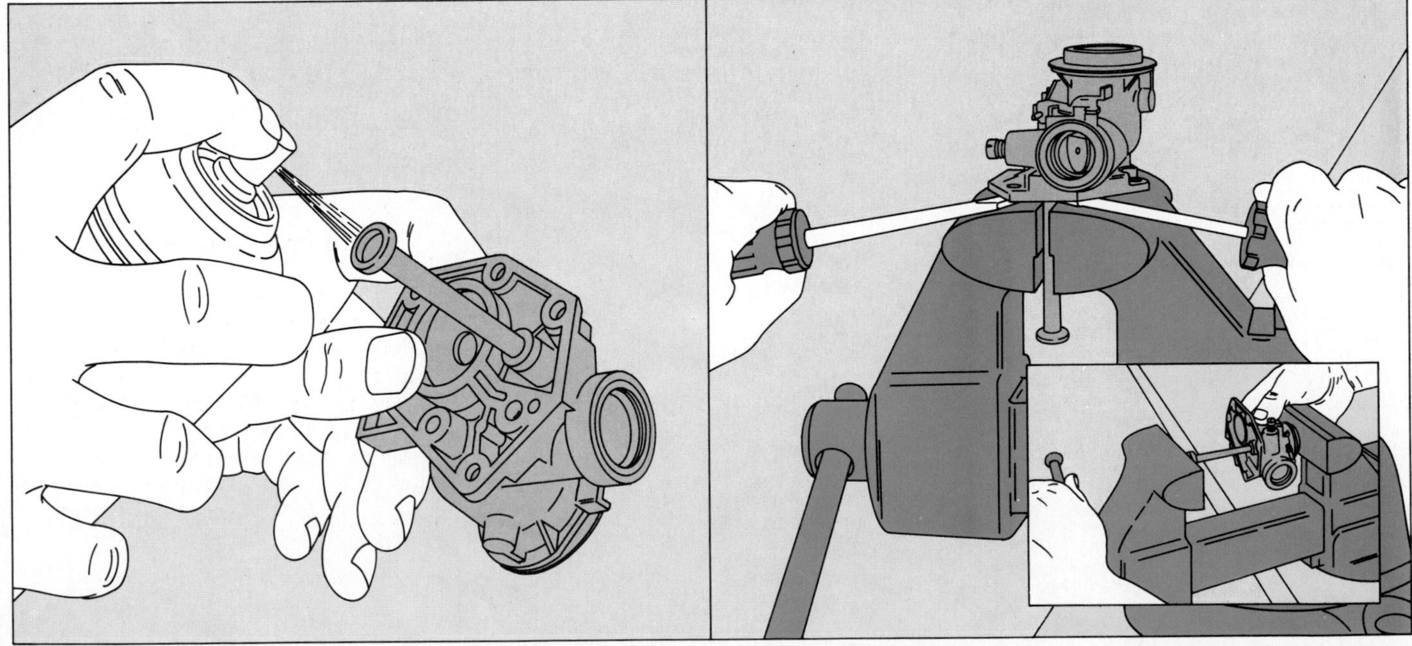

4 Servicing the fuel pipe. Use a straight pin to pry the screen out of the recess at the end of the fuel pipe; then, with the same pin, poke the ball-shaped check valve inside the fuel pipe to see if it moves freely. If the ball is stuck, spray a penetrating lubricant into the pipe *(above, left)*; wait a few minutes, then work the ball loose with the pin. If the fuel pipe cannot be unclogged or if there are kinks in its walls, it should be replaced. Measure and note the distance that the pipe protrudes from its bore in the bottom of the carburetor, then clamp the fuel pipe between the jaws of a bench vise and use two screwdrivers as levers to pry up the carburetor until it can be lifted off the pipe *(above, right)*. To insert a replacement pipe, push the pipe partway into its bore, then hold both the pipe and the carburetor between the wide-open jaws of the vise *(inset)*. Gradually tighten the jaws until the pipe protrudes the same distance as the old pipe.

If the carburetor has a second, nylon fuel pipe with a hexagonal head, remove that pipe with a wrench before the work described above.

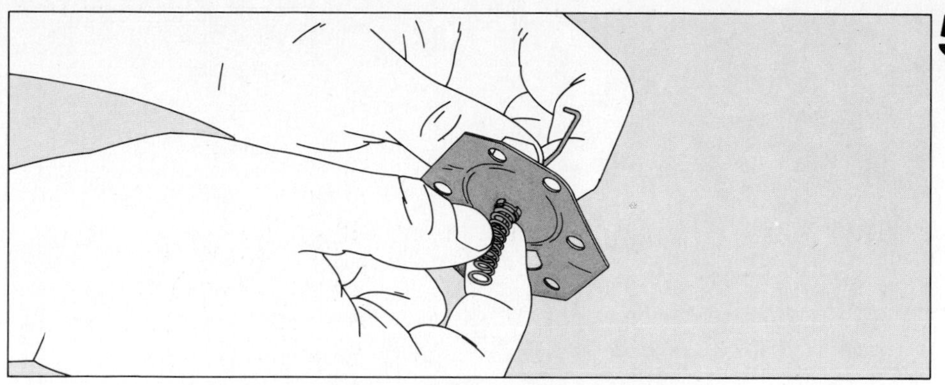

5 **Reassembling the carburetor.** Press the end of the choke spring into place between the flanges on the bottom of the new diaphragm. Then slide the diaphragm over the fuel pipe, aligning its holes with the holes on the carburetor, and push the end of the choke link into its bore *(Step 3, opposite)*. Replace all of the gaskets, and screw the access cover back over the choke linkage. Check the throttle plate for easy movement as on page 64. Remount the carburetor on the fuel tank, and bolt the assembly back onto the engine. Reconnect the governor linkages and the fuel line. Adjust the carburetor screws as illustrated on page 64.

Intricacies of a Pressure-feed Carburetor

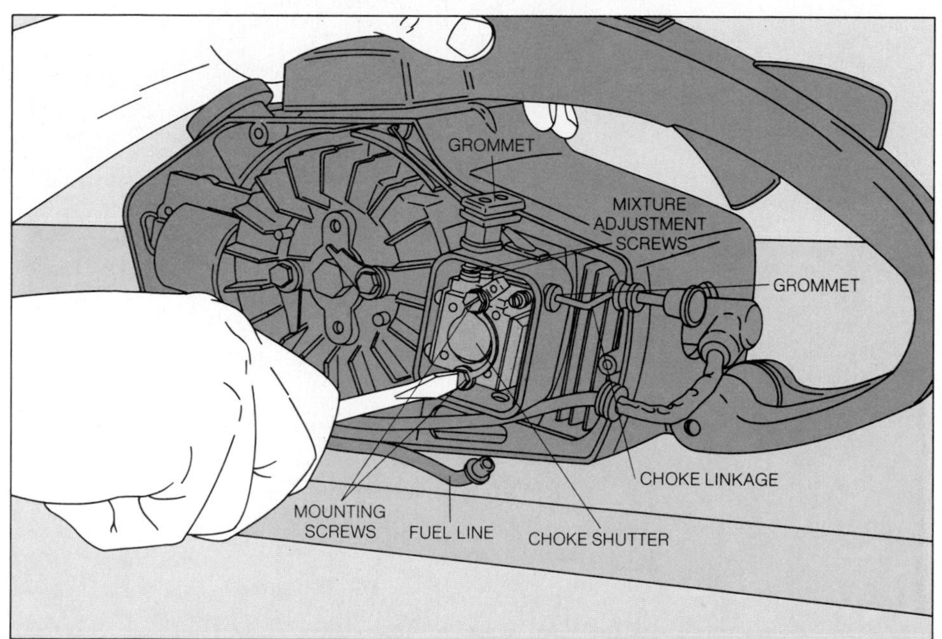

1 **Removing the choke linkage.** After removing the fan housing or air-filter cover, disconnect the fuel line from the carburetor. Use a screwdriver or an open-end wrench to loosen the mounting screws. Pull the screws and choke shutter off the carburetor, at the same time disengaging the choke linkage. Pull the choke linkage and its grommet away from the carburetor housing. Pull the rubber grommet—if there is one—off the tops of the mixture adjustment screws.

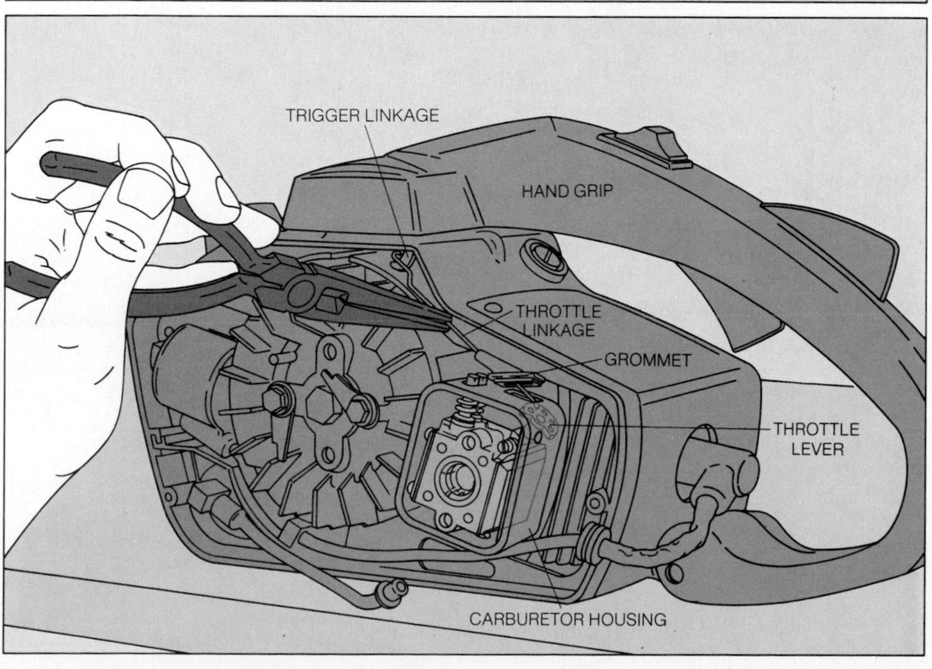

2 **Disengaging the throttle linkage.** Pry the rubber grommet on the throttle linkage from the top of the carburetor housing, then use long-nose pliers to unhook the throttle linkage from the trigger linkage. (On some machines, you may have to remove a portion of the rear hand grip in order to reach the connection.) Unhook the throttle linkage from the throttle lever, noting the position of the link on the lever in order to reconnect it correctly. Remove the carburetor from its housing.

3 Replacing the filter screen. Remove the screw that holds the pump-section cover to the top of the carburetor. Rap the cover lightly around the edges with a plastic-tipped mallet to loosen it. Carefully slip the tip of a screwdriver beneath the cover and its gasket, and gently pry them off. Carefully peel the pump-section diaphragm away from the body of the carburetor *(inset)*. (On some carburetors the metering diaphragm may lie beneath the pump diaphragm, with a thick plastic gasket sandwiched between the two; remove the whole assembly.) Pry the filter screen from its recess with a straight pin, and poke the replacement screen from the rebuilding kit into the hole *(page 62, Step 6)*.

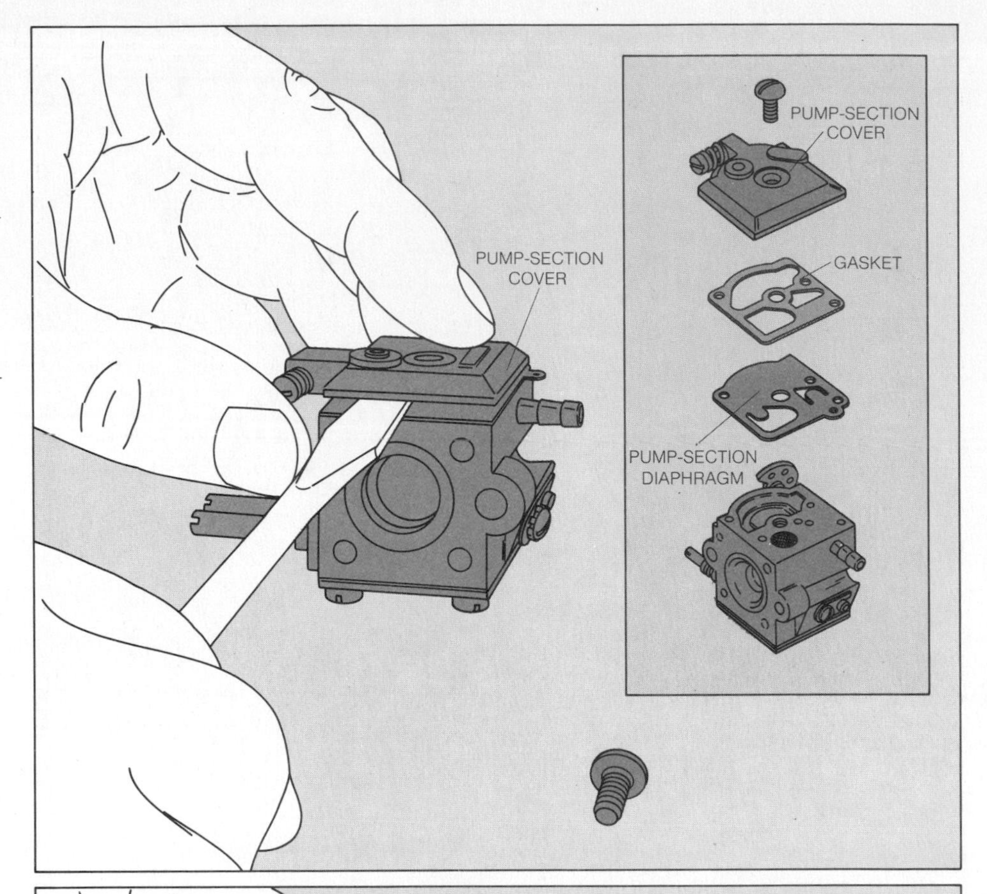

4 Removing the metering diaphragm. Invert the carburetor and remove the metering-diaphragm cover. Carefully lift off the diaphragm, noting how it is connected to the needle-valve assembly. On some models, such as the one pictured here, a tab on the diaphragm presses against a rocker arm; on others, the diaphragm may be hooked directly onto the needle valve.

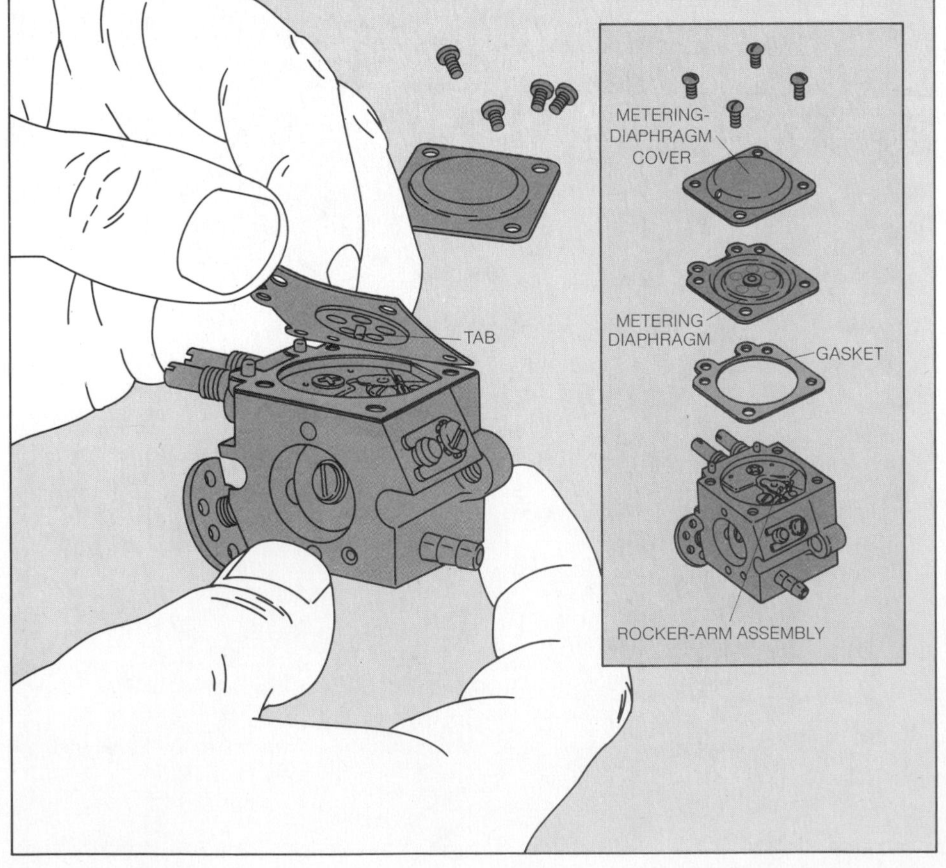

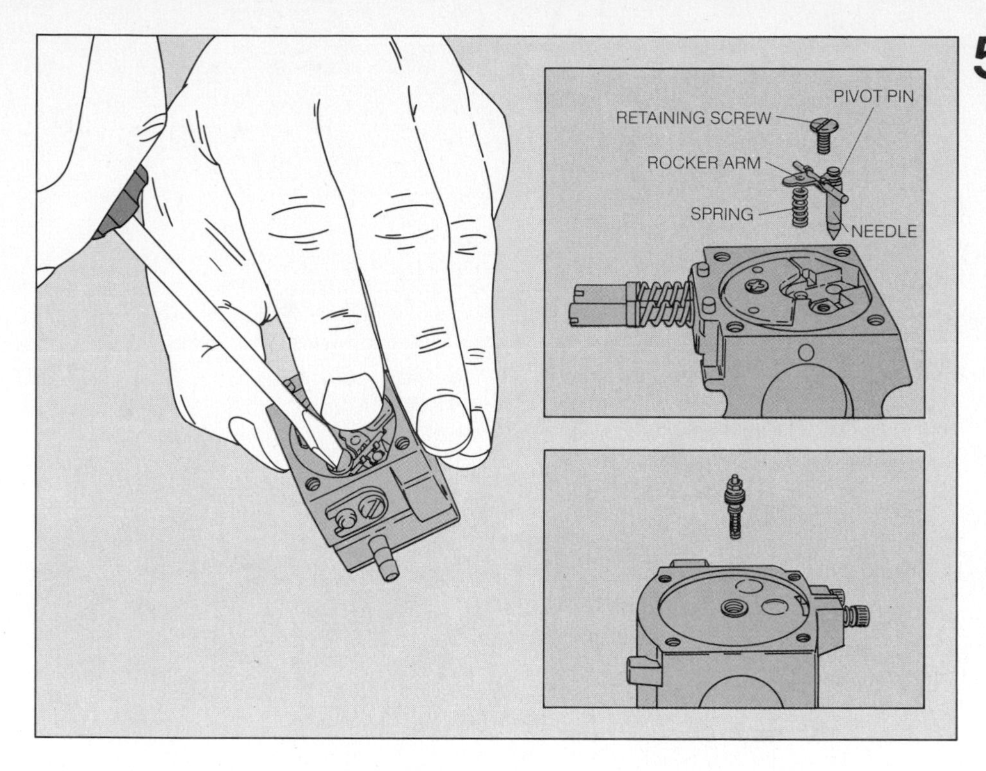

5 Removing the needle-valve assembly. Pressing on the rocker arm to prevent the spring beneath it from suddenly popping up, remove the retaining screw that holds the pivot pin in place. Lift out all of the pieces *(top inset)* and inspect them. If any are worn or damaged, replace them with matching parts from the rebuilding kit.

If the carburetor has the type of diaphragm that presses against a hex-headed, spring-loaded needle valve *(bottom inset)*, use a small socket wrench to loosen the hex nut. Inspect and, if necessary, replace the parts.

Strip the carburetor of any remaining gaskets or nylon parts, then soak and clean it as on page 62. Reassemble it starting with the needle-valve assembly. Hook the notch at the top of the needle over the forked end of the rocker arm and lower the needle into its bore. Fit the spring beneath the rocker arm and tighten the retaining screw. Reattach the diaphragms, gaskets and covers. Reinstall the carburetor, making the preliminary mixture and idle-speed adjustments as shown on page 64.

The Governors That Regulate an Engine's Speed

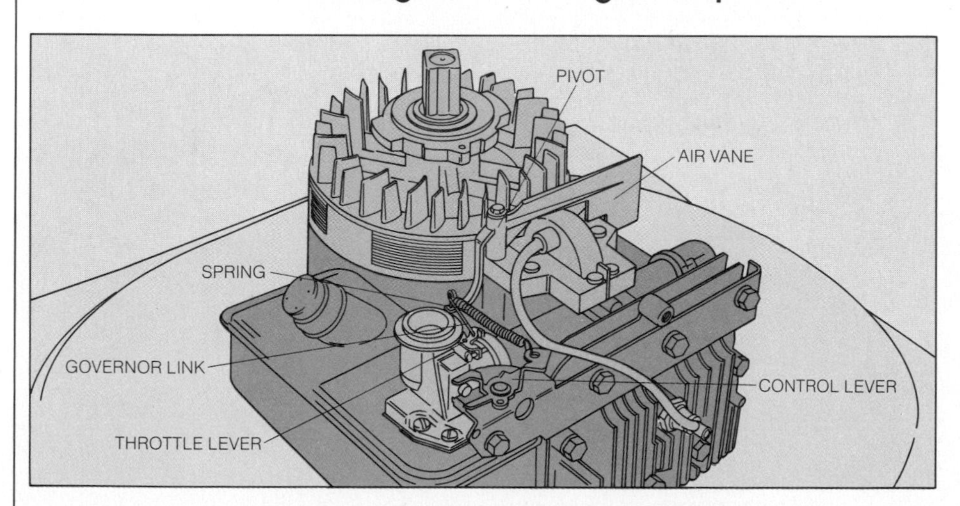

Small engines are designed to operate at smooth, constant speeds: Excessive speed or abrupt changes in speed can jolt engine parts and set up vibrations that cause premature wear. Because gasoline-powered machines are subjected to rapidly shifting work loads that can affect engine speed, they are equipped with governors that make instantaneous throttle adjustments to keep the engines operating evenly.

When a lawnmower, for example, runs across a patch of low grass, its blade speeds up, spinning the engine; when it hits a thicket of tough weeds, the blade and engine slow down. A governor monitors such changes, restricting the flow of fuel to a racing engine and supplying additional fuel to a straining one.

There are two types of small-engine governors. The so-called mechanical governor is operated by centrifugal force. Flyweights, attached to a revolving gear connected to the camshaft, pivot outward when the engine runs too fast, raising a lever that partially closes the throttle. When the engine slows, the weights move inward, opening the throttle.

The air-vane governor, pictured here, is operated by a lightweight metal or plastic arm mounted next to the flywheel. When the engine speeds, the flywheel fins set up a strong breeze that blows on the vane. The vane moves, tugging the governor linkage, which in turn pulls on the throttle lever, partially closing the throttle. If the engine slows down, the flow of air from the flywheel decreases and the vane drifts in the opposite direction, opening the throttle.

Governor settings are meticulously calibrated, and fine tuning is best left to a professional mechanic. But a few simple maintenance chores will keep a governor functioning smoothly. Periodically clean off any dirt that might obstruct the air vane, throttle lever or control lever. Wipe off the pivot and spray it with a penetrating lubricant. Finally, inspect the spring. If it is stretched, kinked or damaged in any way, replace it with a new one designed for the engine.

New Life for Cylinder, Valves and Crankshaft

If your small engine has developed chronic problems, and a tune-up and thorough cleaning of the carburetor have proved bootless, a complete overhaul may be called for. An overhaul is not routine maintenance, but rather the ultimate surgery, resorted to only when all minor adjustments have failed. It consists of a major rebuilding of the engine's innards—the cylinder, piston, rings, valves and crankshaft—and is seldom needed until one or more vital engine parts have plainly come to the end of their days.

The need for an overhaul may be indicated by many symptoms—most often loss of power, excessive oil consumption and hard starting *(pages 32-37)*. But since these same symptoms can be the result of trivial problems as well—a loose spark plug, for example, can cause a severe loss of power—be sure to eliminate all lesser problems as possibilities before deciding that an overhaul is necessary.

To prepare for an overhaul, remove the engine from the tool and drain it of gasoline and oil. Then strip the engine down to the block, removing in turn the engine housing, starter, ignition system *(pages 39-42)* and carburetor *(pages 58-69)*. To keep track of the myriad engine parts in the course of a major disassembly, set them aside in logical groupings, using numbered or colored tape tabs to identify connecting parts, especially wires and springs. Label bolts, nuts and washers with the name of the part they came from, or tape them to the components that they fasten.

When you are disassembling engine-block components, remove each part, clean it thoroughly with parts-cleaning solvent, examine it for defects and broken fasteners, then measure it for excessive wear. There are reasons for adhering strictly to the order of these steps: A part cannot be properly inspected if it is not cleaned first, and time spent measuring a part can be saved if prior inspection has already shown it to be defective. Examining parts in the order shown on the following pages will also save you work: For example, if you discover that the cylinder is worn and must be rebored to a larger size, there is no need to examine the piston and rings—they must be replaced.

If yours is a two-stroke engine, some steps will vary from those shown for a four-stroke engine. Starting on page 81 are special procedures for overhauling two-stroke engines.

Much of this disassembly and measuring may be done with general-purpose mechanic's tools—a socket-wrench set, micrometer, telescoping gauge, feeler gauge and the like. Some overhaul tasks call for special tools *(pages 8-13)*, such as a piston-ring expander for removing and replacing piston rings, a hone for polishing the cylinder walls and a valve-lapping tool for refurbishing valves. None of these tools is expensive, and all are available from auto-supply stores and small-engine shops.

The most important tool of all, however, is one that few people even think of as a tool. A service manual, written specifically for your particular engine model, is an absolute must for an overhaul. Available from the manufacturer, the service manual lists critical measurements for all parts of the engine subject to wear. By comparing your measurements to those listed, you can determine whether parts are so worn that they must be replaced or resurfaced.

Replacing parts is easy. A new crankshaft, piston, rings or valves may be purchased from a factory-authorized dealer. Resurfacing, on the other hand—also called regrinding—is a precision job best left to an authorized service shop equipped with the necessary machining tools. Components that most commonly require professional machining include the cylinder *(page 74)*, the main crankshaft bearings, and the valve seats, faces and guides *(opposite)*.

Replacement parts and machine-shop service, however, are expensive. And if your engine requires much of it, you face a tough decision: Before buying replacement parts for an overhaul or leaving major parts at a machine shop, compare the total cost of repair to the cost of a new engine or tool. If your engine is old and requires much service, it will be more economical in the long run to buy a new one.

If you decide to buy new parts for an overhaul, be sure they are manufactured specifically for your engine make and model. Very few engine parts are interchangeable. Also, when reassembling an engine, pay special attention to the orientation of the parts to each other and to the crankcase. Some parts that look symmetrical, such as pistons and connecting-rod caps, are in fact asymmetrical. These parts are marked with guide notches, arrows or letters—explained by the service manual—to help you replace them correctly. Just before reassembling the engine, clean all of the parts with fresh parts-cleaning solvent, and coat them with a thin film of motor oil.

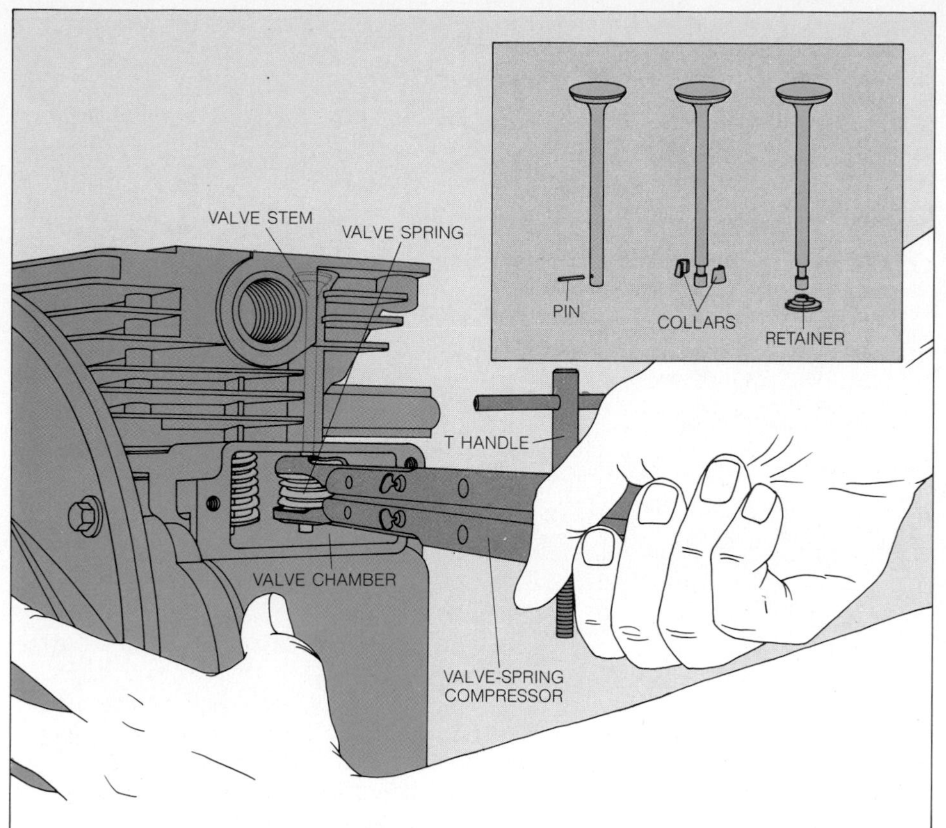

Measuring Valves for a Perfect Fit

1 Removing the valve springs. Take off the cylinder head and breather cover *(pages 47-48)*. Insert a valve-spring compressor into the valve chamber, wedging its upper jaw between the top of the spring and the top of the chamber and its lower jaw beneath the valve retainer at the bottom of the spring. Turn the T handle of the tool to close the jaws, compressing the spring.

If the valve is held in place by a pin or a pair of collars *(inset)*, dislodge them with your finger or a screwdriver after compressing the spring. If the valve has a slotted retainer, push the retainer to snap the slot off the valve stem. Lift the valve out of its guide. Withdraw the compressor from the valve chamber, loosen the T handle to release the spring, and set the parts aside.

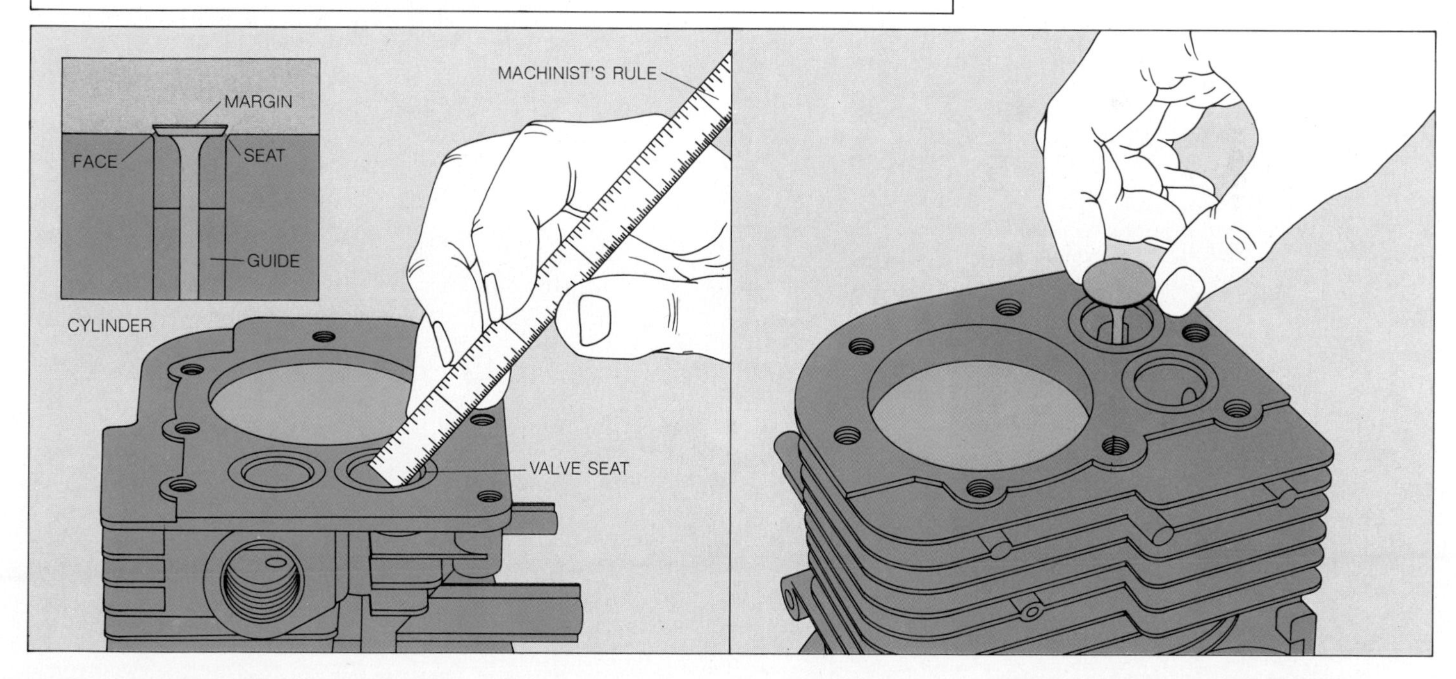

2 Inspecting the valves. Before measuring the valve seats, clean the valves and the seats carefully, scraping off carbon deposits with an old screwdriver or a wire brush. Inspect the valves and seats for melted areas, cracks or obvious, severe wear. Check for cracks in the engine block between the valve seat and the cylinder. To determine whether the valve stems are bent, roll the valves on a table, with the valve heads overhanging the table edge—if a stem is bent, the valve will wobble as it rolls. Any of these defects means replacement of the faulty part.

With a machinist's rule *(above, left)*, measure the parts of the valves and seats; compare the measurements of the seats, faces and margins *(inset)* to the measurements in the service manual. If the valve seat is too wide or the valve margin too narrow, replace the valves or have them reground by a machinist. Insert each valve in its guide; if you can see that the valve seat and valve do not seal completely, have them reground. Lift each valve ½ inch *(above, right)*, and wiggle the stem in the guide; if you feel more than a tiny bit of play, have the guide refurbished by a machinist, who will ream it out and insert a bushing to meet the right specification.

Inspecting the Camshaft and Crankshaft

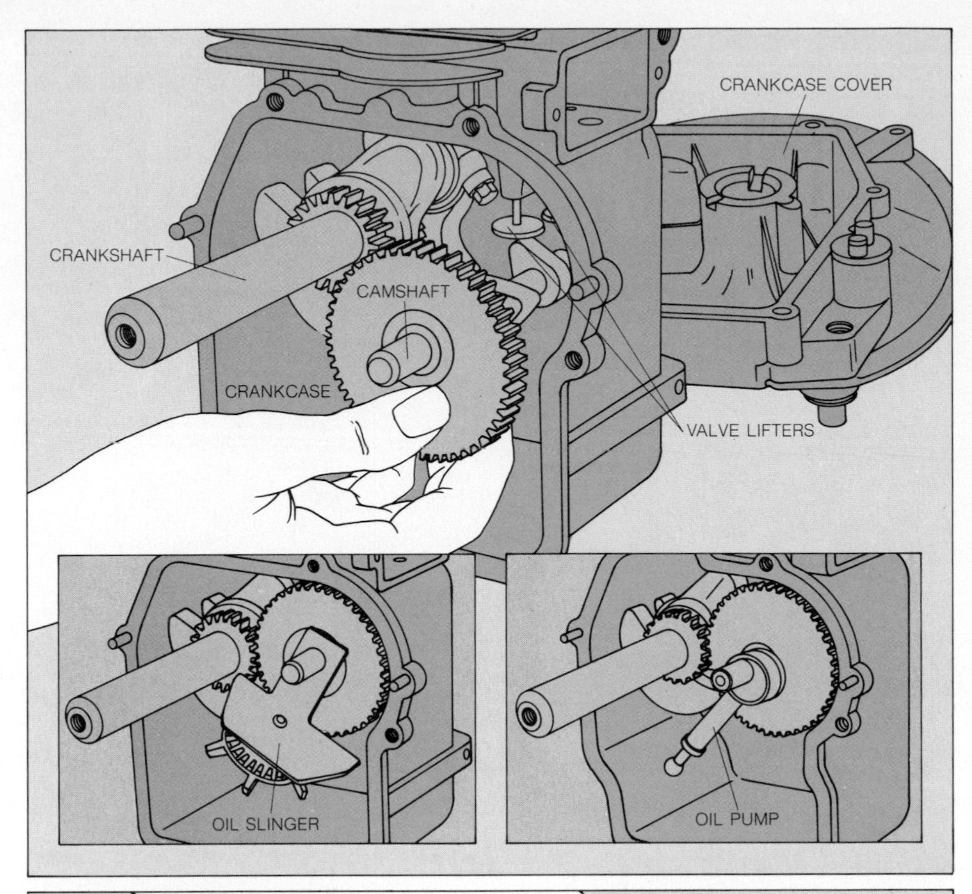

1 Removing the camshaft. After servicing the valves and valve springs, clean the dirt and rust off the exposed end of the crankshaft with emery cloth. Remove the cover to expose the camshaft gear. Grasp the gear wheel, and pull it and the camshaft straight out. The valve lifters—small, plunger-like devices that raise and lower the valves—may fall out of the valve guides into the crankcase; if they do not, pull them out.

On some engines, an oil slinger *(left inset)* or an oil pump *(right inset)* will be attached to the front of the cam gear. Lift the device off the camshaft—you may have to pry loose a small clip to do so—and inspect it to see that it operates smoothly. If it binds or if you see other obvious damage such as worn gear teeth or nicks, replace the part.

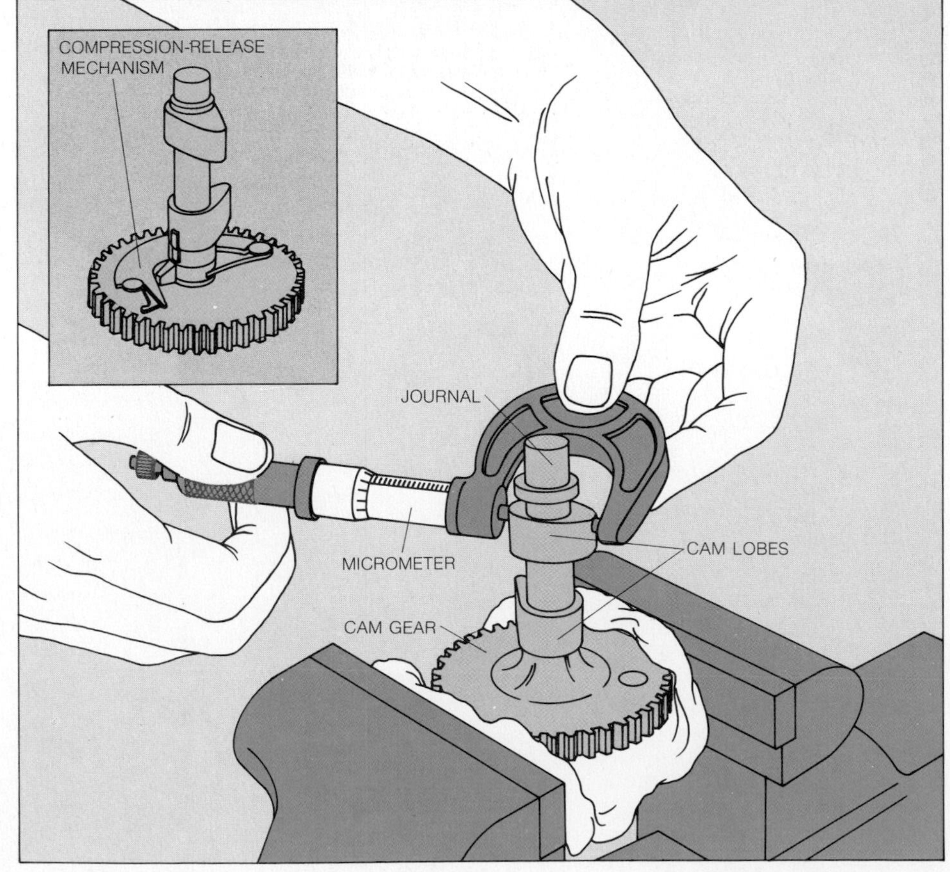

2 Inspecting the camshaft. Clean the camshaft with parts-cleaning solvent. Examine the camshaft gear teeth for nicks or wear and the camshaft journals—the smoothed, end sections of the shaft that ride in supporting bearings in the walls of the crankcase—for deep scratches or gouges. Replace the camshaft if you see these defects. Place the cam gear in a vise, using a rag to protect the teeth, and measure the cams—your service manual may call them the lobes—and journals with a micrometer *(page 13)*. Compare the measurements to the reject, or minimum allowable, sizes in the service manual. If any part is worn smaller than this size, the entire shaft must be replaced.

Check the back of the cam gear for a compression-release mechanism *(inset);* some engines require them for easier starting. Flip the mechanism away from the camshaft with your finger and let it snap back. If the mechanism binds or is sluggish, replace it.

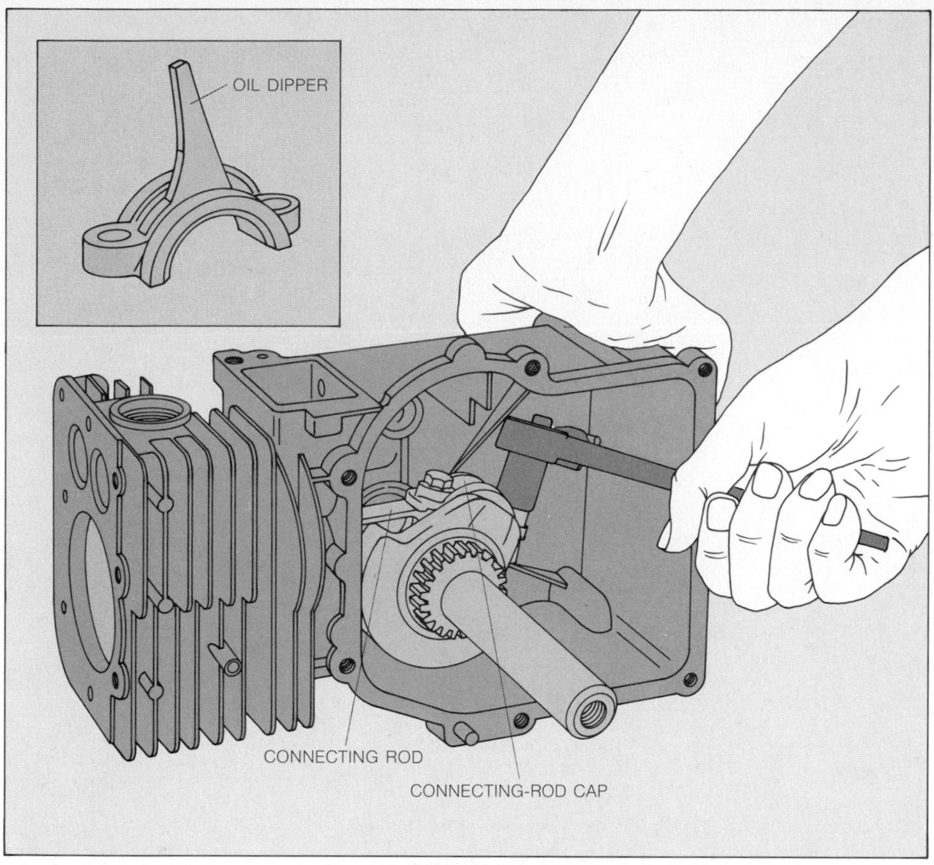

3 Releasing the connecting rod. Turn the crankshaft by hand until you can reach the connecting-rod cap, which secures the rod to the crankshaft; turning the engine block on its side, as here, may make the cap easier to reach. If the cap has a locking plate *(page 79, inset)*, bend down its tabs. Using a socket wrench, unscrew the fasteners that hold the cap to the connecting rod, and pull off the cap. Examine the polished inside bearing surface of the cap for scratches or gouges. If the cap—or any other part of the connecting-rod-and-cap assembly—is damaged, the entire assembly must be replaced so that all of the parts wear evenly. Note: Some connecting-rod caps differ from the type shown here, with an oil dipper attached to the cap to facilitate lubrication *(inset);* all caps, however, are removed and inspected in the same way.

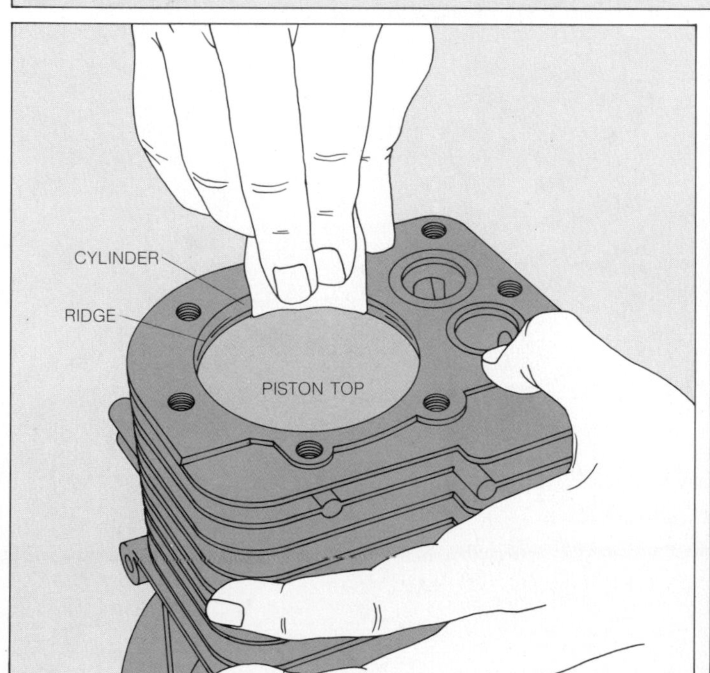

4 Freeing the piston. A ridge of carbon and unworn metal, which forms on the cylinder wall just above the piston, must be removed before you can push the piston out. Set the engine block upright. Turn the crankshaft to raise the piston to the ridge. Rub the ridge flat with a folded piece of emery cloth, taking care not to scratch the top rim of the cylinder wall.

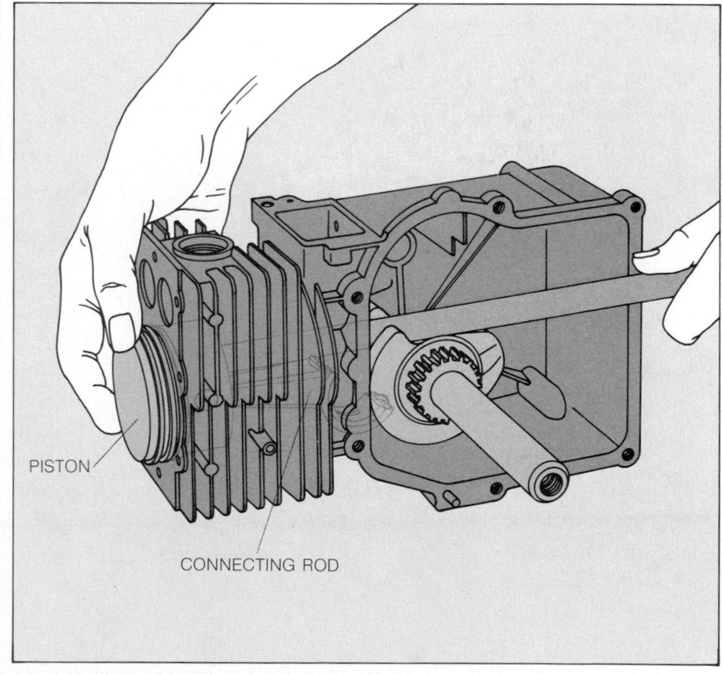

5 Pushing out the piston. Place the engine on its side, and turn the crankshaft so that you can fit the end of a wooden dowel under the piston, alongside the connecting rod, or against the end of the connecting rod. Push the dowel to force the piston and connecting rod out of the cylinder *(inset)*. Then pull the crankshaft straight out of the crankcase.

6 **Examining the crankshaft.** Clean the crankshaft with solvent, and inspect it for obvious defects: cracks, bending, worn gear teeth, damaged flywheel threads, bent flywheel keyways or scored journals. A deep blue cast to the metal, called bluing, indicates a special kind of damage. It is evidence that the metal has overheated and weakened. If you find any of these defects, replace the crankshaft. If not, measure the diameters of the connecting-rod, power-take-off and flywheel journals with a micrometer and compare them with the minimum allowable sizes in the service manual; if any journal is worn too small, replace the crankshaft.

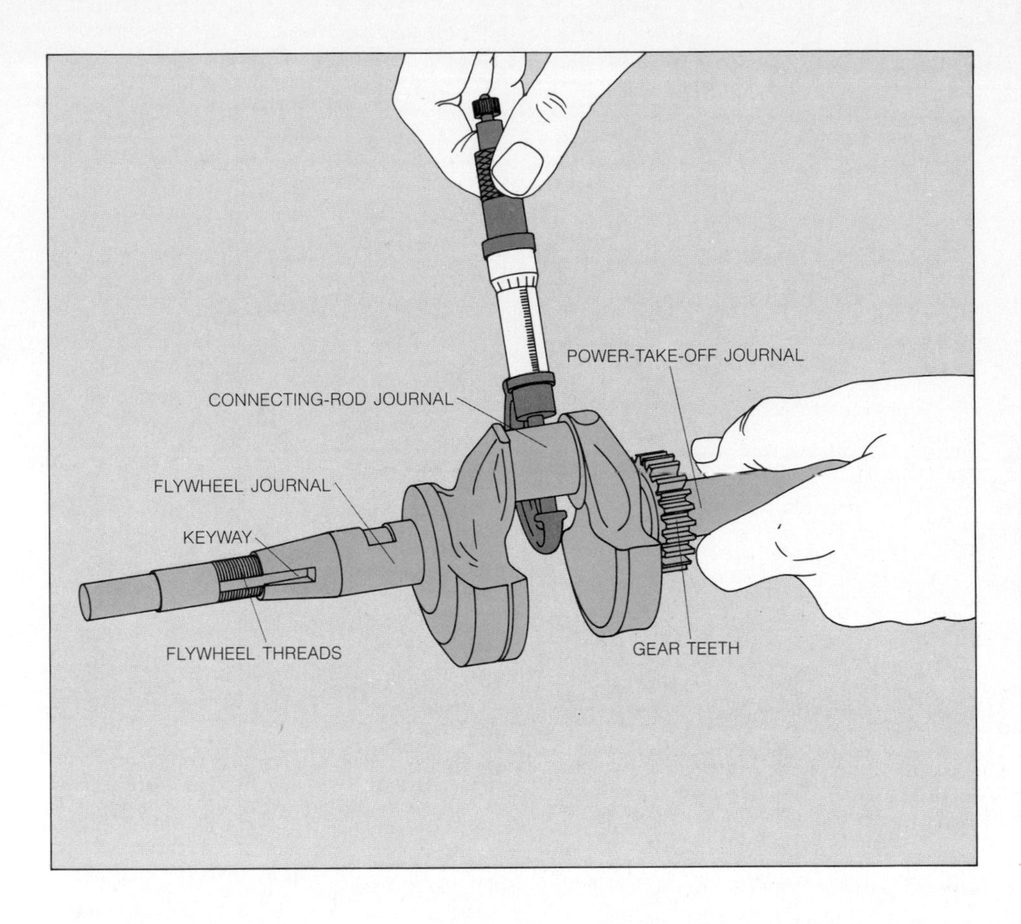

Testing the Cylinder, Piston and Rings

1 **Evaluating the cylinder.** Clean the inside of the cylinder and inspect it for cracks, bluing and scoring. Then use a telescoping gauge *(page 13)* to measure the inside diameter at three different depths, turning the gauge 90° at each depth to check the cylinder's roundness. Compare all of the measurements with the specifications in the service manual to determine if the cylinder has worn too far out of round or beyond the maximum allowable diameter, then use your findings to decide what to do next: No remedy exists for cracks or bluing; either one indicates that the cylinder is beyond repair and must be replaced. Scoring or excessive wear, however, can be remedied: Many cylinders can be bored larger by a machinist and then fitted with an oversized piston and set of piston rings. Some cylinders cannot be rebored; the service manual will indicate whether yours must be replaced instead.

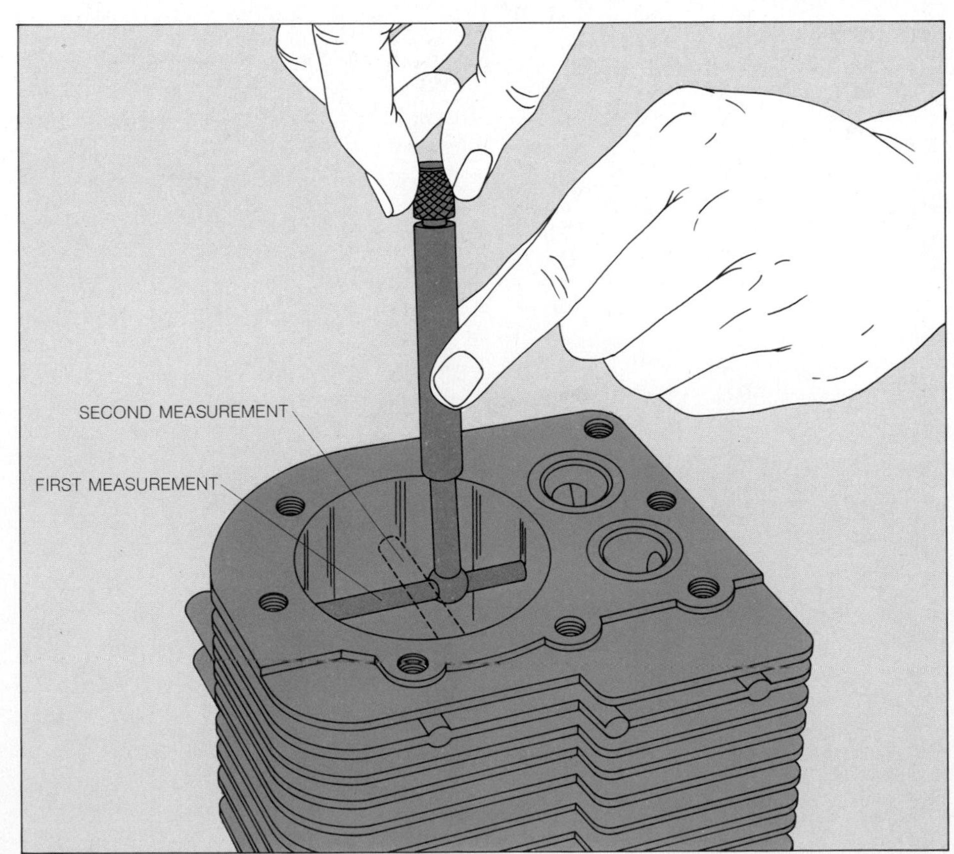

2 Freeing the connecting rod. To remove the wrist pin—the small spindle that joins the connecting rod and the piston—grasp the clip at each end of the pin with long-nose pliers *(below, left)*. Drive out the pin with a wooden dowel *(below, right)*, lightly tapping the dowel on a table if necessary. No longer held by the wrist pin, the connecting rod will fall out of the piston. Inspect the wrist pin for scoring and bluing.
Look for scores on the bearing surface of the connecting rod, and check that the rod is straight. If you see a defect, replace the rod and the cap.

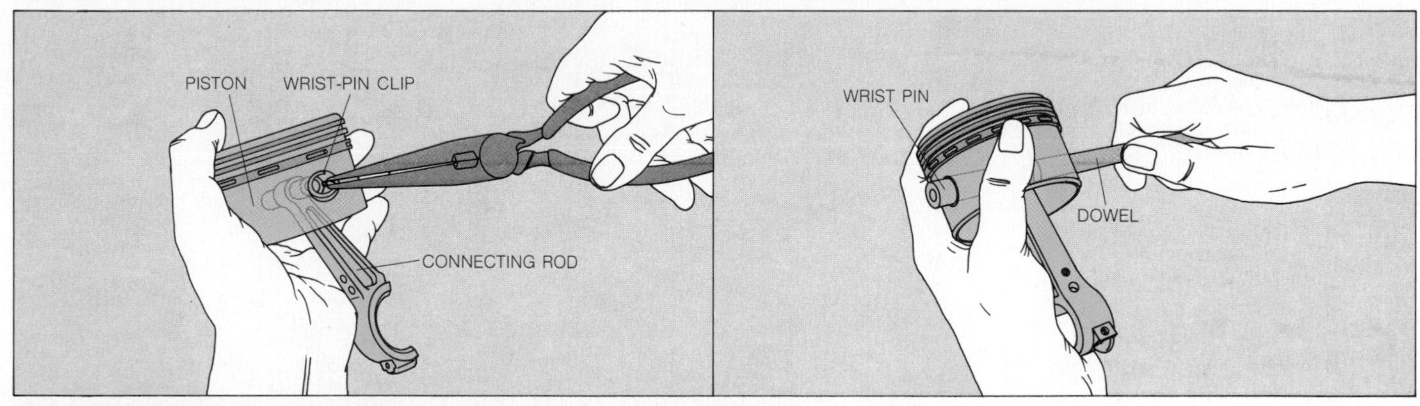

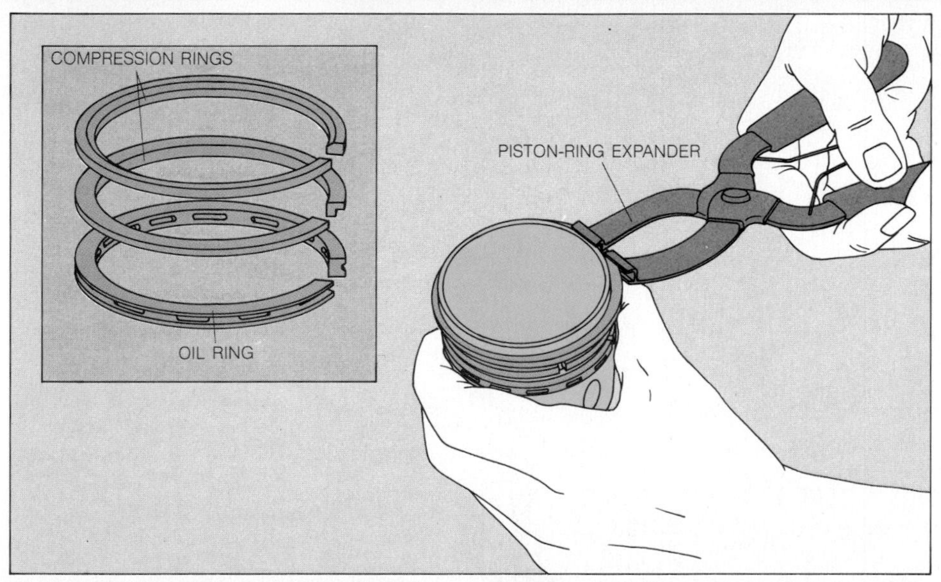

3 Removing the piston rings. Engage the slotted jaws of a piston-ring expander in the gap of the top piston ring. Gently squeeze the tool, forcing the jaws apart to enlarge the ring. Slip the ring off the piston. In the same way, remove the remaining two rings, noting their order: The two upper rings are built to seal in compression; the oil ring below them keeps oil out of the combustion chamber. Each has a distinctive profile *(inset)* and fits only one groove in the piston. Clean the rings and examine them for nicks and fissures. All three rings must be replaced if any one of them is defective.

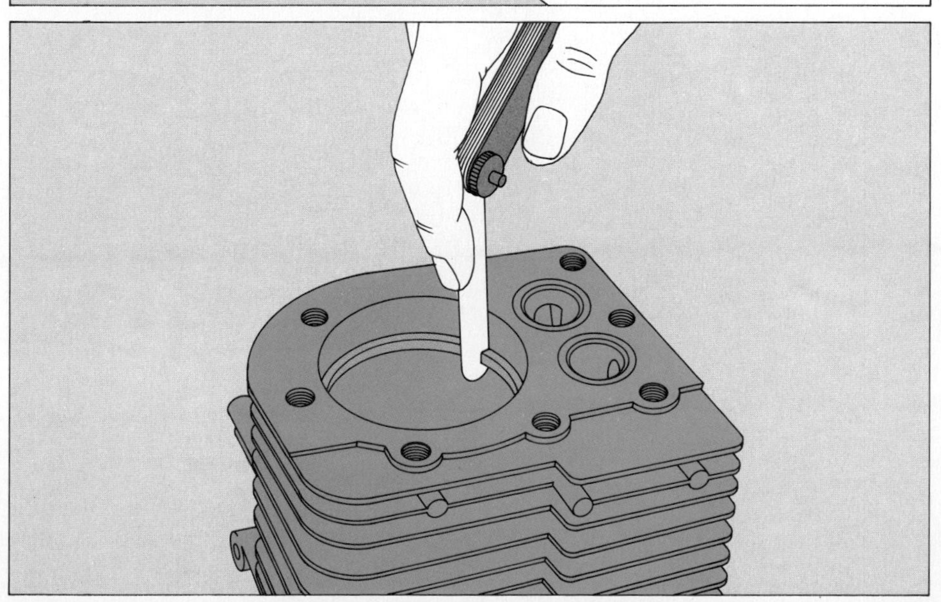

4 Measuring the rings for wear. Squeeze one ring, and set it in the cylinder. Using the top of the piston as a plunger, push the ring down 1 inch. Insert feeler gauges *(page 12)* into the ring gap until you find the largest feeler gauge—or combination of gauges—that will slide in and out. This is the ring-gap size. Similarly measure the gap of the remaining two rings. If the measurement for any ring is larger than the maximum specified in the service manual, the ring is worn and all three rings must be replaced.

5 Cleaning the piston. If you must replace the piston rings, break an old ring in half and use its factory-cut (unbroken) end to scrape debris from the ring grooves of the piston. If you are not discarding the rings, use an old small screwdriver as a scraper. In either case, work gently to avoid scratching off any metal. Dissolve any stubborn deposits with parts-cleaning solvent and, finally, wipe the grooves with a rag. Inspect the cleaned piston for scoring, cracks, or burning of the piston top and uppermost ring land—the flat part of the piston's side between ring grooves. Any of these defects requires replacement of the piston and rings.

6 Measuring ring-groove clearance. Using a piston-ring expander, reinstall the old piston rings if they passed the ring-gap test *(Step 4)*, or install new rings. Slide feeler gauges into the groove under each ring to measure the clearance; if the clearance of any ring exceeds the maximum specified in the service manual, discard the old rings and repeat the test with new ones. If the clearance is too great with new rings, the ring grooves on the piston are worn out and the piston must be replaced.

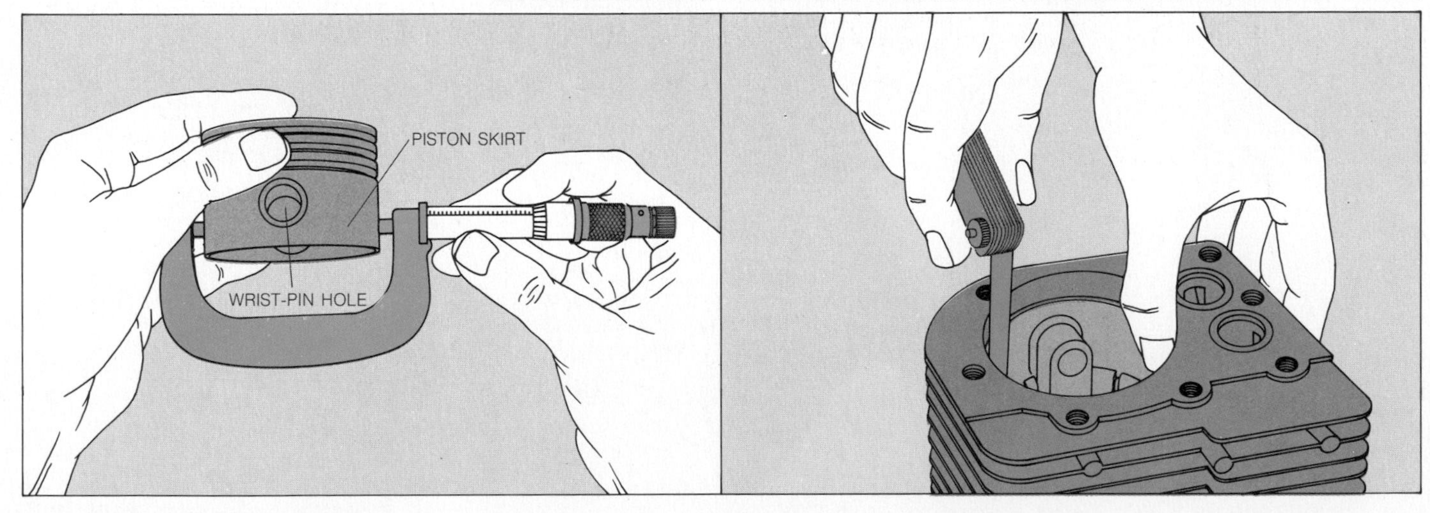

7 Measuring the fit of piston and cylinder. The tiny clearance between the piston and the cylinder is measured in one of two ways: If your service manual gives the diameter of the piston skirt—the smooth part of the piston below the wrist-pin holes—simply measure it with a micrometer *(above, left)*. Since pistons are often made out of round, however, be sure to measure the skirt at the points specified by the service manual. If the piston-skirt diameter and the cylinder diameter *(Step 1)* are both within the specified ranges, then the clearance between piston and cylinder is normal. If the piston is worn too small, replace it and the rings.

If your service manual gives the piston-to-cylinder clearance, measure the clearance with feeler gauges *(above, right)*. Hold a feeler gauge against the piston skirt, 90° away from the wrist-pin holes; insert the feeler gauge and the piston, with rings removed, upside down into the cylinder. Withdraw the gauge, feeling for moderate drag *(page 12)*. If necessary, reinsert the piston with increasingly larger gauges until you find the largest one that fits. Compare the clearance to that specified in the service manual. Replace the piston and rings if the clearance is too large.

Refinishing the Cylinder and Valves

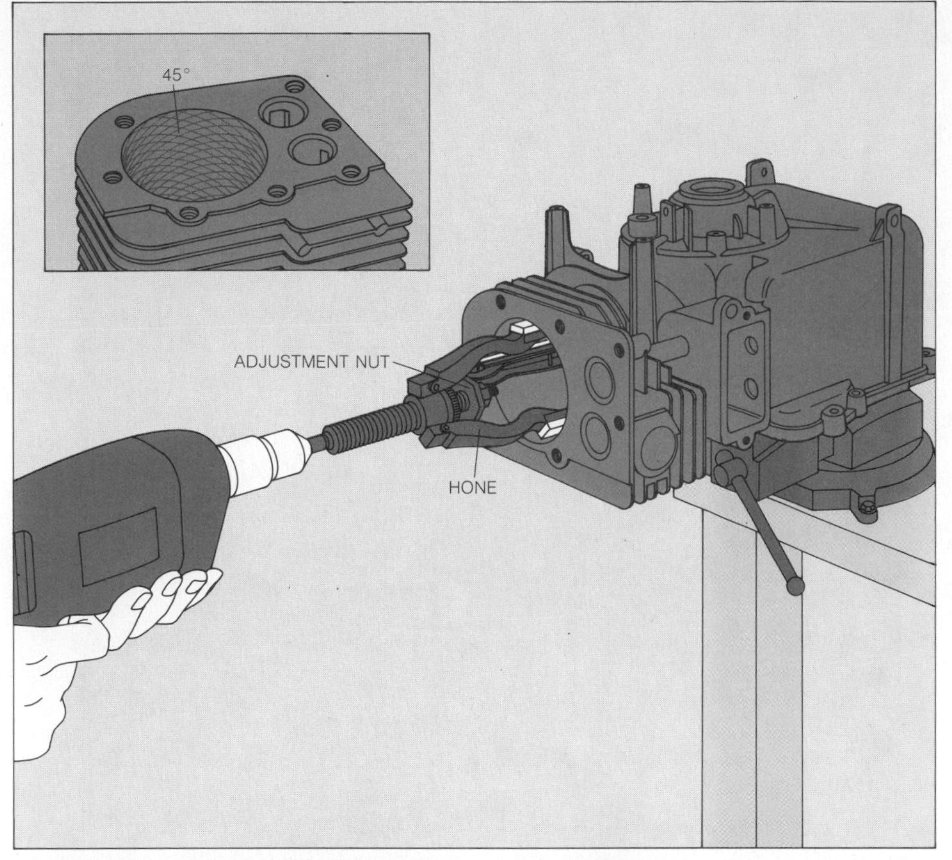

Deglazing the cylinder. Secure the engine block in a vise padded with rags or wood blocks. Insert a cylinder hone *(page 9)*, fitted with fine-grade cutting stones, into the cylinder. Tighten the hone-adjustment nut to adjust the pressure of the stones against the cylinder. Spray the stones and cylinder with light household oil. Fit an electric drill to the protruding shank of the hone, and run the drill at low speed—from 300 to 700 rpm's—moving the hone back and forth inside the cylinder about once a second. Do not push or pull the stones more than ¾ inch out of either end of the cylinder.

After about 10 seconds, stop the drill to check for a 45° crosshatch pattern on the cylinder wall *(inset)*. If the pattern is circular, move the drill in and out faster; if the lines run lengthwise along the cylinder, slide the drill more slowly. If the hone-adjustment nut loosens, retighten it and continue deglazing. To avoid wearing away the cylinder too much, do not run the hone for more than three minutes and stop as soon as the cylinder has a uniformly crosshatched surface. Clean the cylinder wall with parts-cleaning solvent, and coat it with motor oil.

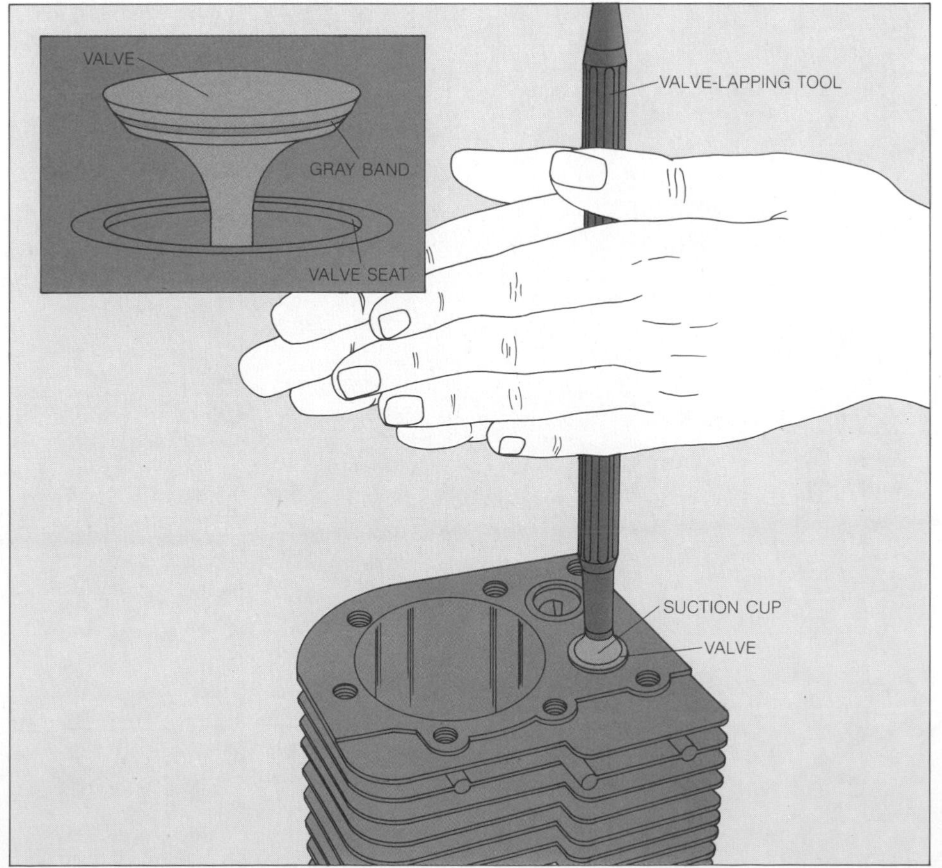

Polishing the valves. After any routine inspection of the valve seats *(page 71, Step 2)* or after having the valves reground, you must lap—or polish—the valves to perfect their seal. To do this, spread valve-grinding compound—an abrasive paste—on the face of the valve and insert the valve into its guide. Then press the suction cup of a valve-lapping tool onto the head of the valve, and roll the tool back and forth between your hands while pressing downward. Lift the valve at frequent intervals to check the seat and face; stop polishing when an even gray band as wide as the valve seat extends all around the valve face *(inset)*.

Use parts-cleaning solvent to clean away any stray grinding compound, taking care not to wash any of it onto other parts of the engine block.

Reassembling the Engine Step by Step

1 Replacing the oil seals. Ring-shaped oil seals, made of rubber and metal *(inset)*, prevent oil from leaking out of the engine at the main bearings. (In small engines, these bearings are simply smooth holes in the crankcase and crankcase cover, through which the crankshaft protrudes.) Before reinstalling the crankshaft, pry out both old oil seals with a screwdriver, or push them out from the inside. Set the crankcase on its side with the main bearing facing upward. Smear motor oil on a new seal and center it—metal side up—over the bearing. Hold an old socket from a socket-wrench set or a short piece of pipe against the seal; tap it into the bearing with a mallet, leaving the seal flush with the surface of the crankcase or slightly recessed, according to the instructions in your service manual. Replace the remaining oil seal in the main bearing of the crankcase cover, and then place the crankshaft back in the crankcase.

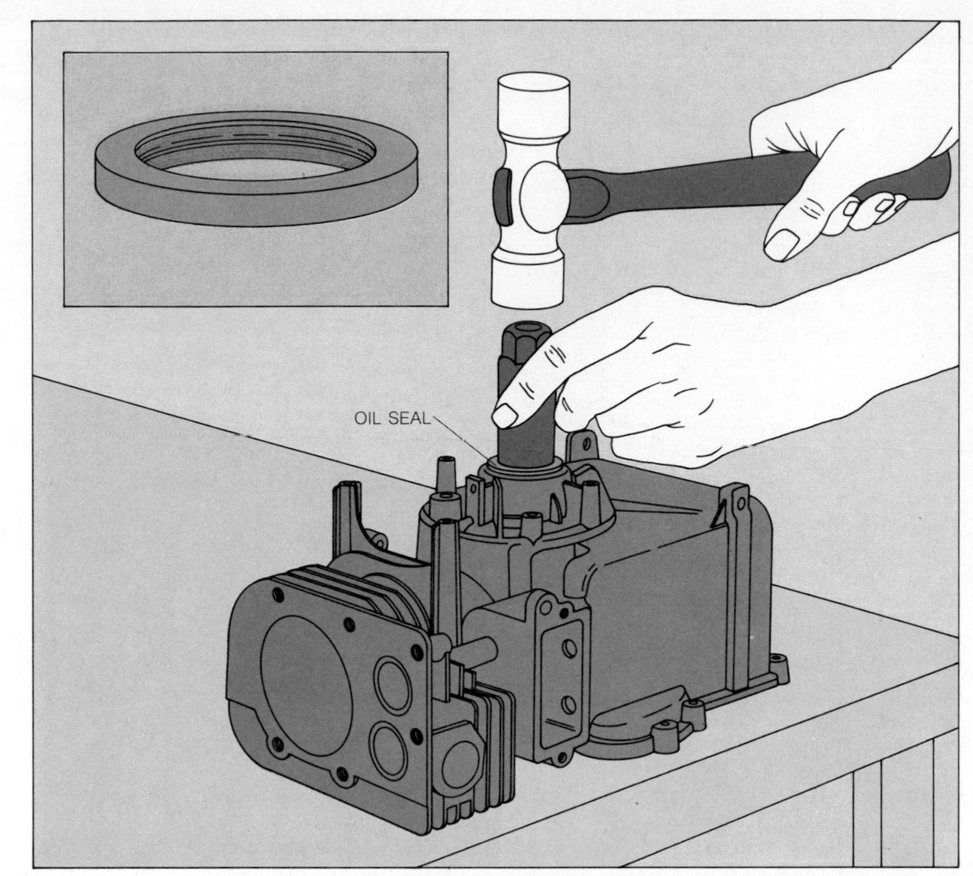

2 Compressing the piston rings. Reassemble the piston, piston rings, wrist pin and connecting rod according to the alignment marks on each part. Stagger the positions of the piston-ring gaps; if they are aligned, the engine can lose compression. Slip a piston-ring compressor or a hose clamp, which you can buy at any auto-supply store, over the head of the piston. Tighten the clamp to compress the piston rings flush with the outside of the piston.

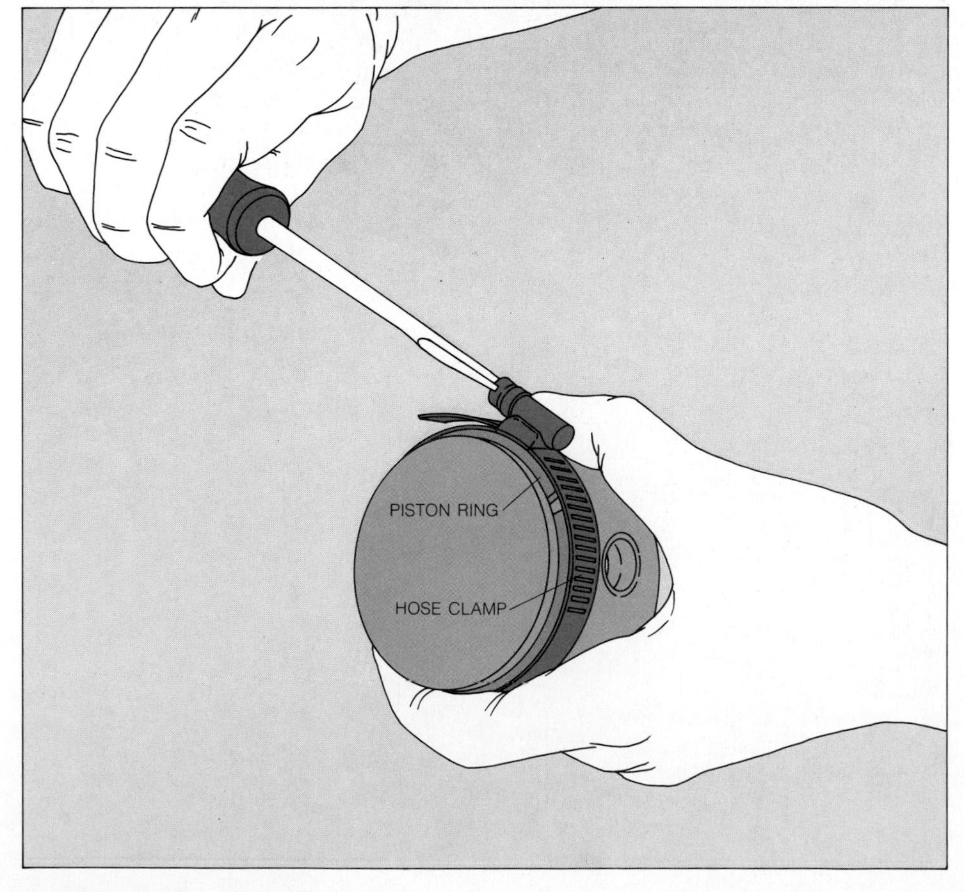

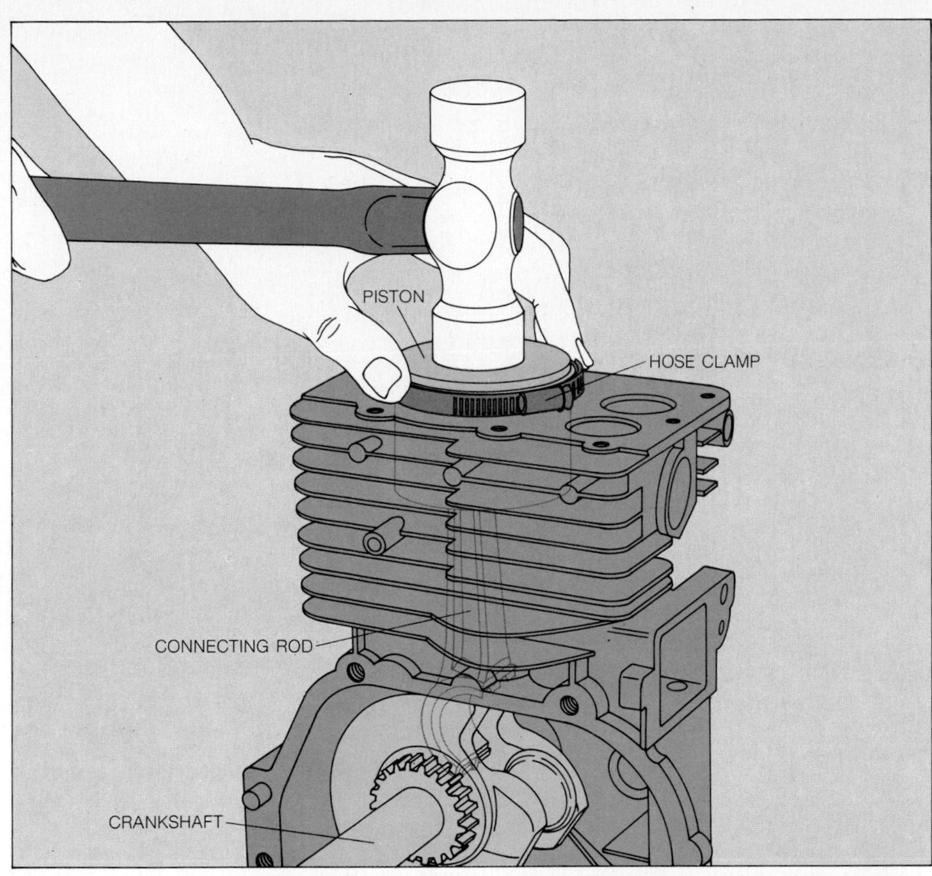

3 Installing the piston in the cylinder. Lubricate the piston, rings and cylinder with motor oil. Insert the connecting rod and piston into the cylinder until the hose clamp stops them. The piston must be inserted so that the alignment mark on the crankshaft end of the connecting rod is facing the crankcase opening. Use a plastic or wooden mallet to tap the piston gently down into the cylinder. If the rings catch on the cylinder edge, tighten the hose clamp; do not force the piston. As the piston slides down, fit the end of the connecting rod onto the crankshaft journal. The hose clamp will slip off as the piston slides into the cylinder.

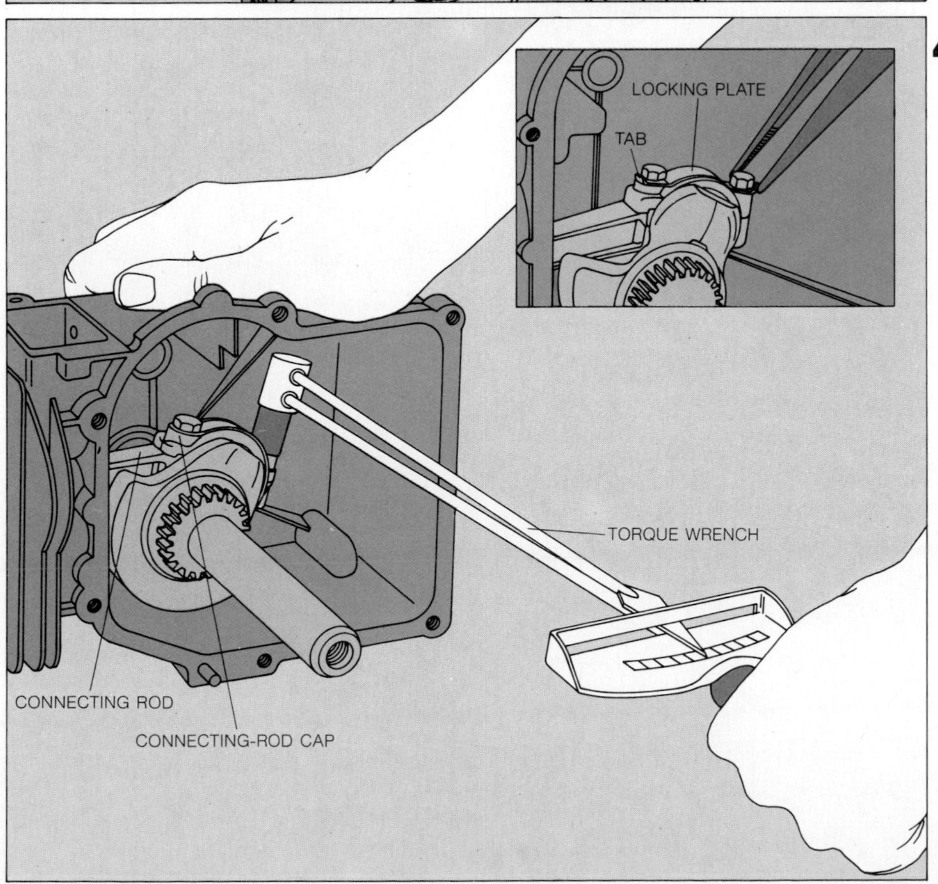

4 Capping the connecting rod. Spread motor oil on the connecting-rod end and cap. Fit the connecting-rod cap onto the end of the connecting rod, enclosing the crankshaft journal. Wiggle the cap if necessary to line up the alignment marks on the rod and the cap, and install the oil dipper *(page 73, Step 3)* if your engine has one. Slip a locking plate—a thin, flat metal strip recommended for preventing critical nuts and bolts from loosening after overhauls—over the cap. Pass the connecting-rod bolts through the holes in the locking plate, hand-tighten the bolts, and then finish tightening them with a torque wrench to the torque specified in the service manual. Bend the tabs of the locking plate up against the flat sides of the bolt heads, using long-nose pliers *(inset)*.

5 **Positioning the camshaft.** After replacing the valve lifters in their guides in the crankcase *(page 72, Step 1)*, turn the crankshaft until the alignment mark on its gear points toward the camshaft position. Insert the camshaft into the crankcase, rotating it so that the alignment mark on its gear meets the one on the crankshaft gear as the gears mesh. Install the oil pump or the oil slinger *(page 72, Step 1)* on the camshaft if the engine has one or the other. Install a new crankcase cover gasket; replace the cover, tightening the bolts in the sequence and to the torque specified by your service manual.

Install the valves, valve springs and retainers, using a valve-spring compressor *(page 71, Step 1)*, and adjust the valve clearance as you would during a tune-up *(page 48)*. Replace the head gasket and the cylinder head as demonstrated on page 48.

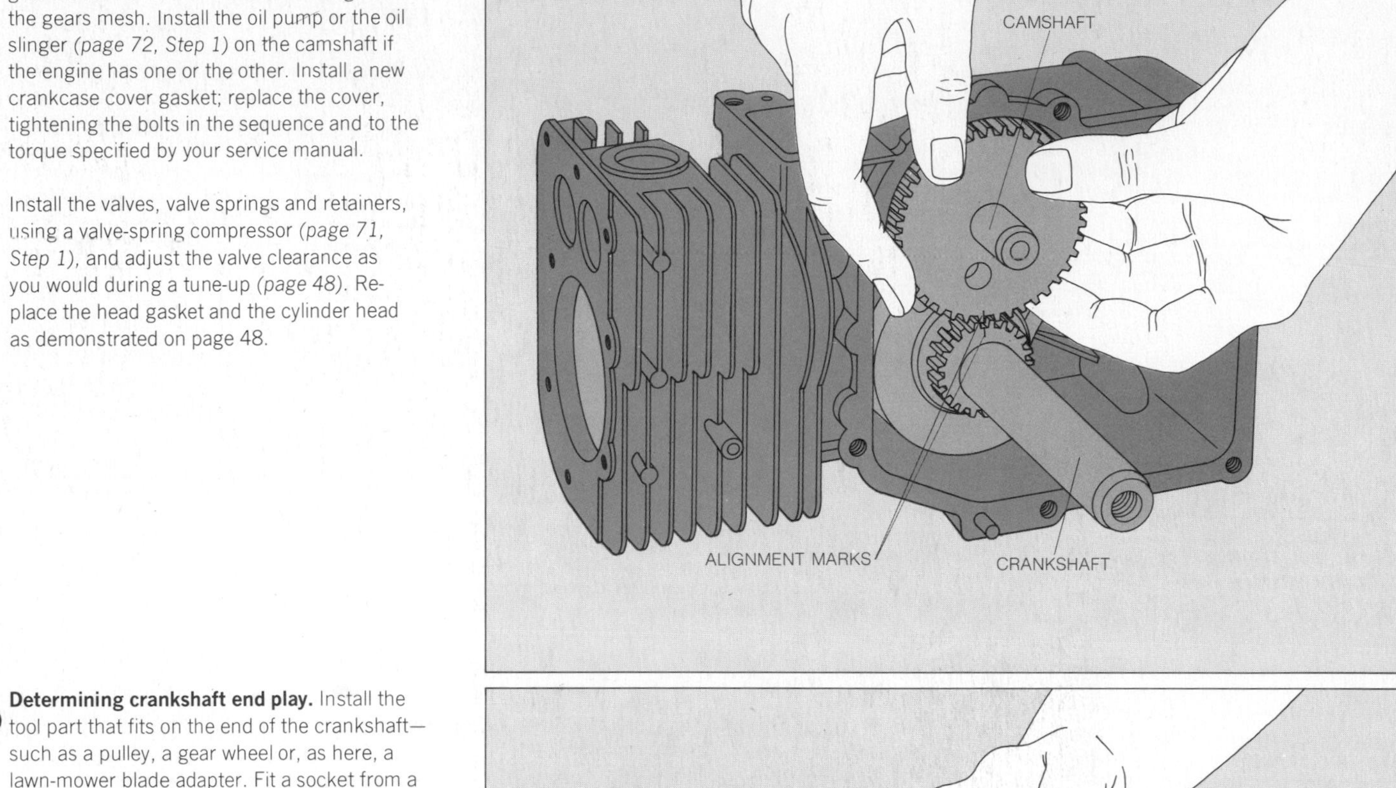

6 **Determining crankshaft end play.** Install the tool part that fits on the end of the crankshaft—such as a pulley, a gear wheel or, as here, a lawn-mower blade adapter. Fit a socket from a socket-wrench set or a length of pipe next to the crankshaft between the part and the crankcase; select a socket or pipe that will leave just enough room for the insertion of a feeler-gauge blade against the crankcase. Slide the crankshaft in and out. Measure the gap with the shaft pushed in all the way and then with the shaft pulled out all the way; the difference between these measurements is the crankshaft end play.

If the end play is incorrect according to the service manual, buy an adjustment kit containing assorted washers and crankcase gaskets: Install a thicker crankcase gasket, or combination of gaskets, to increase end play; add a washer to the crankshaft power-take-off end to decrease end play. A combination of gaskets and washers may be required to produce the recommended end-play measurement.

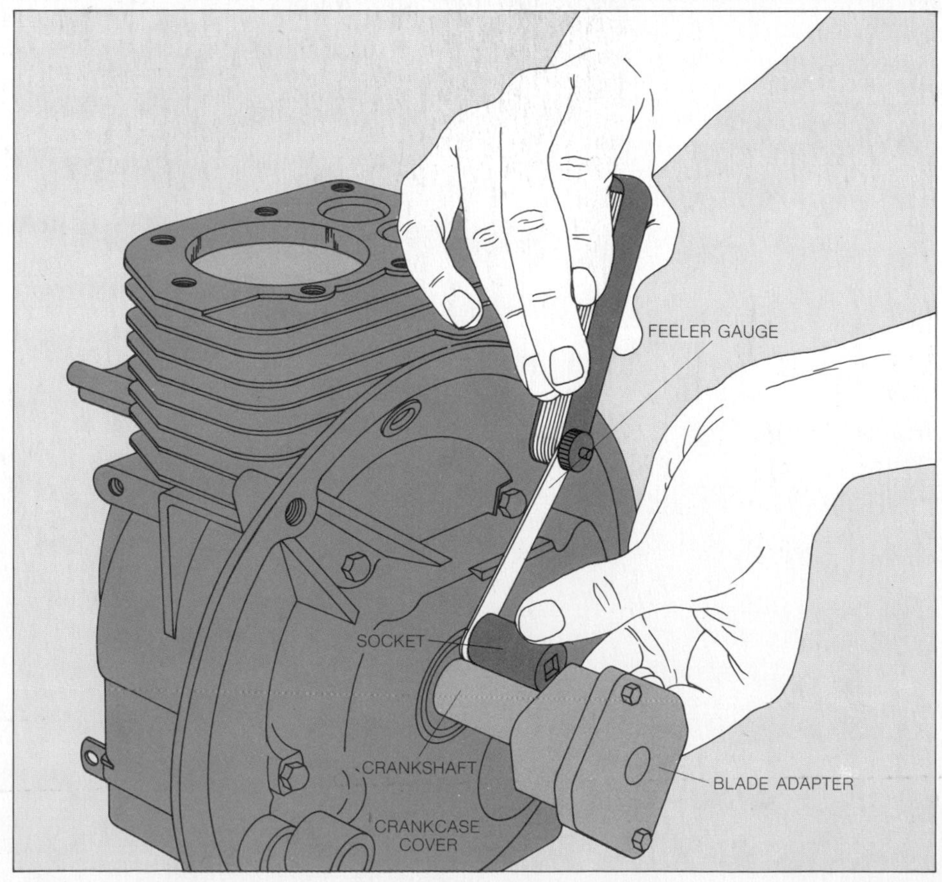

Renovating a Two-stroke

A two-stroke engine is much simpler to overhaul than a four-stroke engine. Because fuel is conveyed to the combustion chamber through ports in the engine block *(page 24)*, there are neither valve surfaces nor a camshaft to service. And since the two-stroke engine is lubricated by oil mixed into the gasoline, there is no oil pump, slinger or dipper to examine.

To overhaul a two-stroke engine, follow the same procedure shown for a four-stroke engine *(pages 70-80)*, but skip steps that do not apply—that is, steps dealing with parts your engine simply does not have. Instead, perform the special jobs shown on the following pages, which relate only to two-stroke engines, and consult your service manual for the peculiarities of your particular model.

When overhauling a two-stroke engine, you will notice some engineering simplifications that make two-stroke engines comparatively light in weight so that they can be used to power hand-held tools, such as chain saws. The tool housing, for example, may double as the crankcase and include the gasoline tank. The cylinder and cylinder head are frequently a single piece that can be unfastened from the crankcase to reveal the piston and the connecting rod, which may not be detachable from the crankshaft. And instead of the bolts and other large fasteners that are employed to hold together a four-stroke engine, you may find small Phillips-head screws or allen screws.

Because two-stroke engines, with their lack of a separate oil reservoir or oil pump, are lubricated in a different way, they require more sophisticated bearings than many four-stroke engines; most are equipped with ball bearings *(page 83)* or needle bearings *(page 82)* to reduce the friction of their rotating parts to an absolute minimum. These bearings are located at critical wearing surfaces: in the holes where the crankshaft rides on the crankcase and at both ends of the connecting rod, where it joins the wrist pin and crankshaft. Examine the bearings carefully in the course of overhauling a two-stroke engine, and replace them if they show signs of wear.

Working on Assemblies of Parts

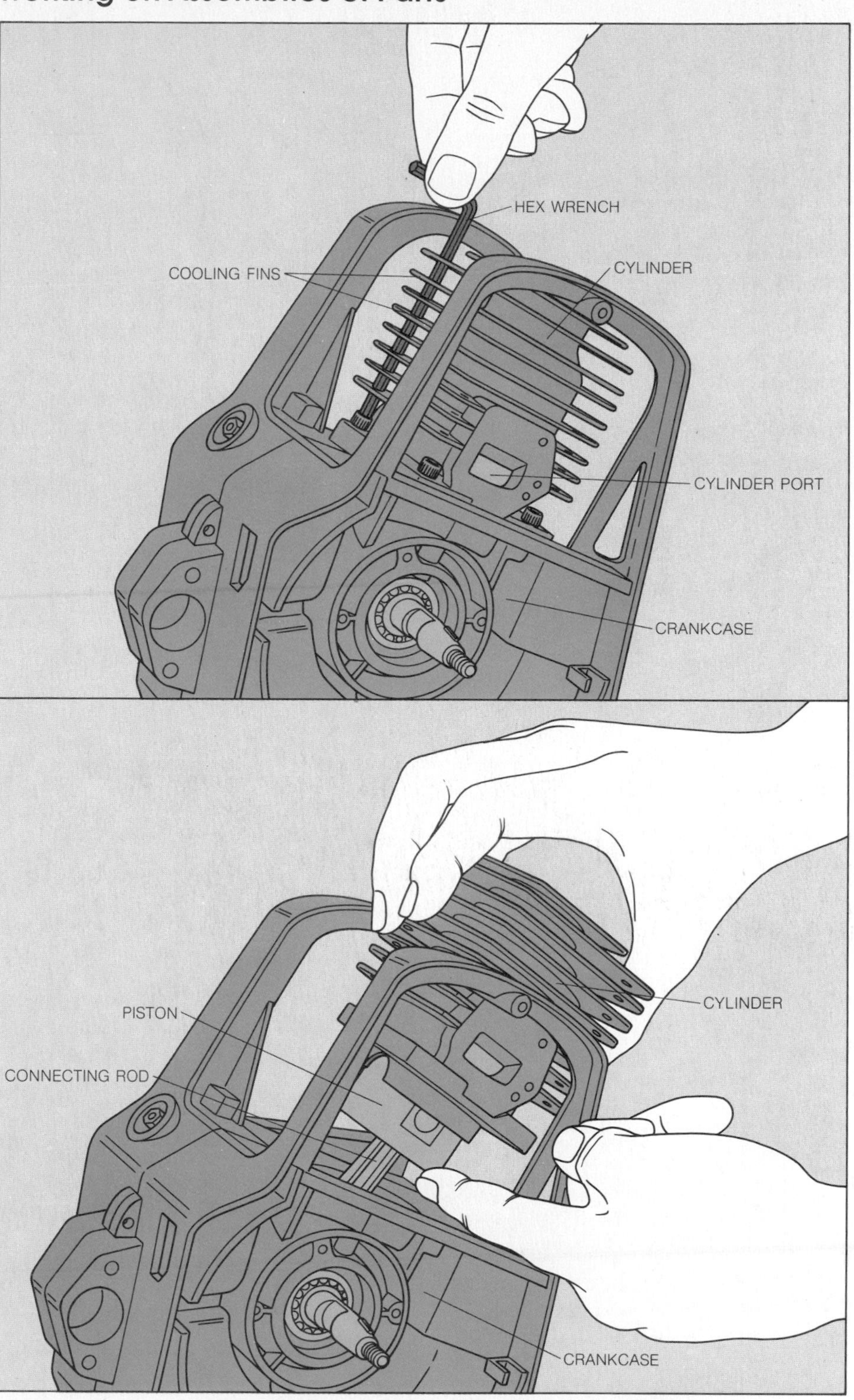

1 Removing the cylinder. Remove external parts such as the muffler and the carburetor, the cylinder head if detachable, and the housing. Insert a hex wrench through the holes in the cylinder's cooling fins *(top)*, and loosen the allen screws that connect cylinder to crankcase. Lift the cylinder *(bottom)* straight up from the crankcase. Clean, inspect and measure the cylinder as for a four-stroke engine *(page 74, Step 1)*; also, feel the beveled edges of the cylinder ports to make sure they are smooth. Replace the cylinder if it is defective. You may also have to replace the piston rings if you replace the cylinder; check the service manual.

2 Removing and inspecting the piston. Detach the piston from the connecting rod by snapping off the wrist-pin clips and pushing out the wrist pin with a wooden dowel, as for a four-stroke engine *(page 75, Step 2)*. Inspect the needle bearing between the wrist pin and the connecting rod *(inset)*. If it shows signs of bluing caused by overheating, is missing needles or does not roll smoothly, replace it.

Inspect and measure the piston's two compression rings *(page 75, Steps 3 and 4)*—two-stroke engines have no oil ring—and examine the piston for wear *(page 76, Steps 5 and 6)*. Finally, check the ring-locator pins, small studs in the piston-ring grooves that prevent the rings from revolving. You must replace the entire piston if the pins are worn, broken or loose. Note the alignment arrow or notch on top of the piston; when the engine is reassembled, this arrow must point toward the cylinder exhaust port.

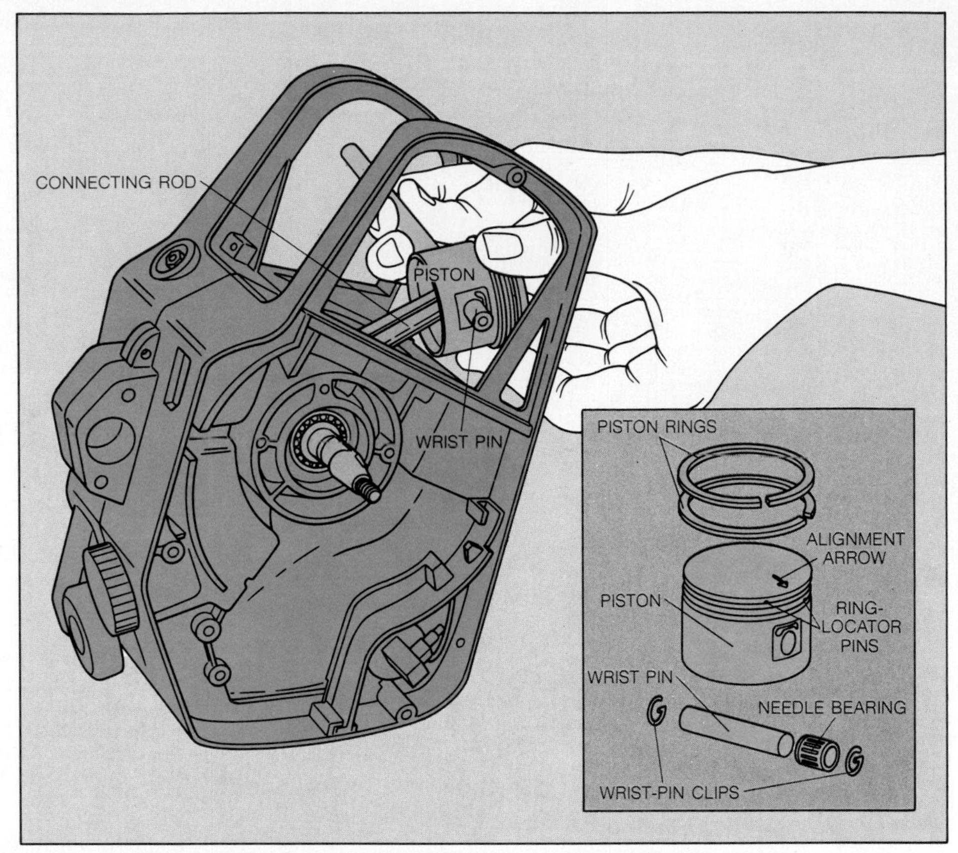

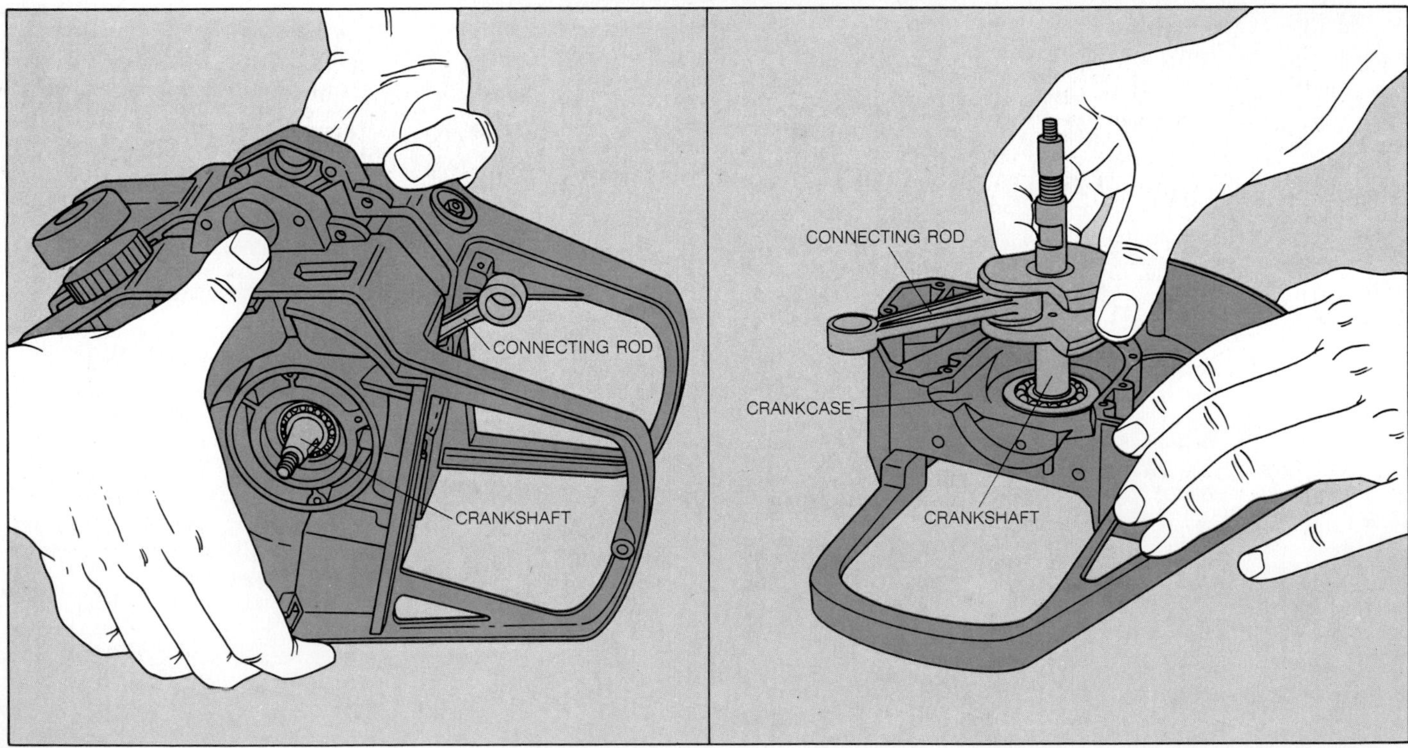

3 Opening the crankcase. Remove the starter *(page 84)*, flywheel and other external equipment as you would with a four-stroke engine. If the crankcase is an inseparable section of the tool housing *(above, left)*, unscrew the two halves of the housing and pull them apart, tapping with a plastic mallet if necessary. If the crankcase is separate from the housing, expose the crankcase by removing the housing, then dismantle the halves of the crankcase. Set the crankcase on its side and lift out the connecting rod and crankshaft assembly *(above, right)*. Inspect both parts for damage, and measure the crankshaft journals for wear *(page 74, Step 6; page 75, Step 2)*.

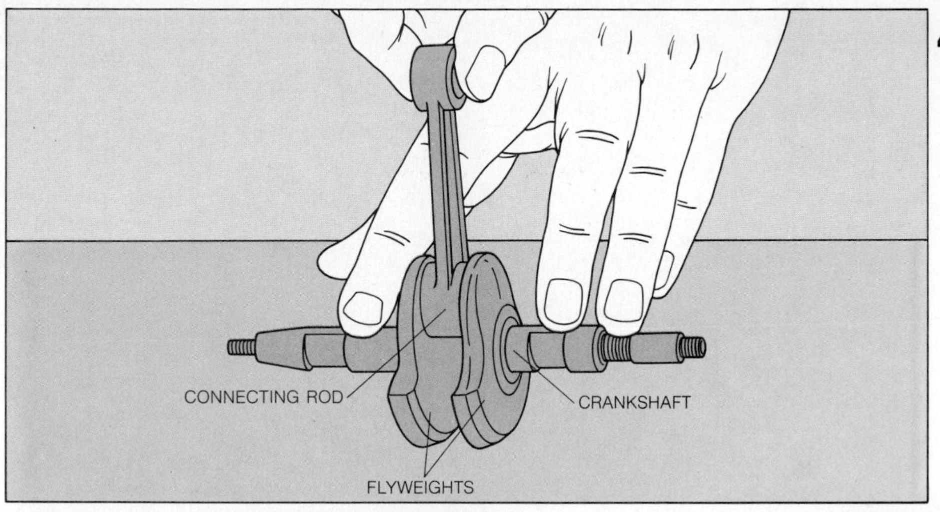

4 Testing connecting-rod play. Hold the crankshaft flyweights against a tabletop with one hand as you wiggle the connecting rod straight up and down with the other hand. If you feel any looseness in the joint between the connecting rod and the crankshaft, the engine assembly must be replaced; a small amount of side-to-side motion, however, is normal.

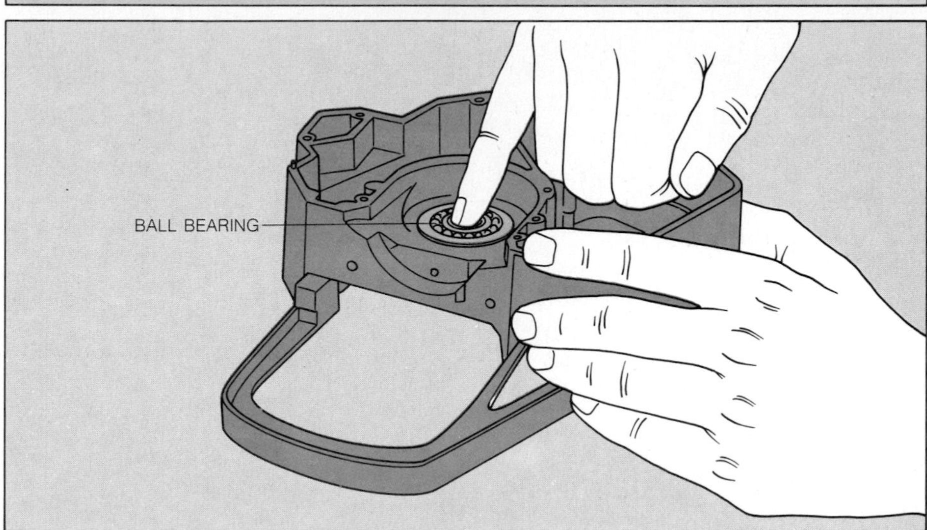

5 Inspecting the crankcase bearings. In two-stroke engines, a ring-shaped ball bearing in each half of the crankcase supports the ends of the crankshaft. Turn each bearing with your finger to check for binding or roughness. Feel the inside surface of the bearing for gouges or scoring, and look for bluing of the metal.

If you detect a defect, replace the bearing. Briefly heat the crankcase around the bearing with a small propane torch to expand the metal encasing the bearing. Tap against the bearing from the outside of the crankcase with a plastic mallet to knock the bearing out. To install a new bearing, heat the crankcase and drop the bearing into place from inside the crankcase.

Servicing a Reed Valve

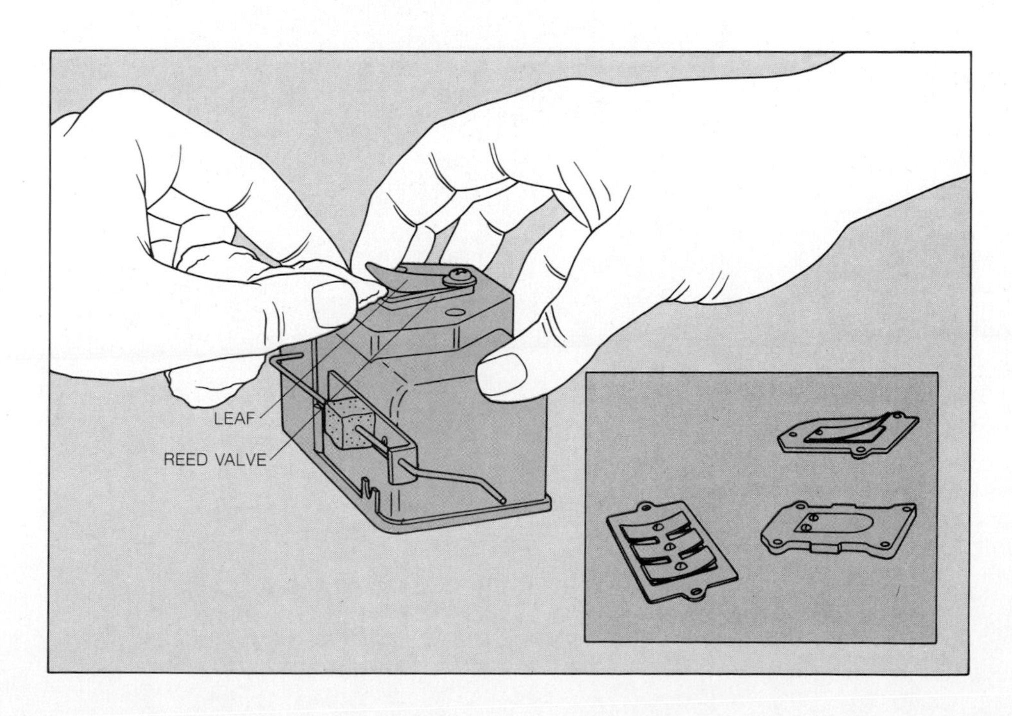

Inspecting a reed valve. The reed valve is a thin metal leaf, sandwiched between the carburetor and the crankcase on most two-stroke engines. The vacuum created by the upward stroke of the piston pulls open the thin leaf to admit fuel to the engine's crankcase *(page 24)*. During the piston's downward stroke, the leaf snaps shut so that engine compression is not lost. Reed valves come in many styles *(inset)*, some with one leaf, others with several.

Clean a reed valve gently, using a soft cloth or a cotton swab to wipe away deposits lodged under the leaf, which may cause the valve to stick open and lose engine compression. To check the operation of the valve, lift the flap with your finger and let go—it should snap shut smartly. If the valve is bent or sticks open, unscrew it from the engine and replace it.

Rope Starters: Simple in Design, Easy to Fix

Rope starters, whether you pull the rope horizontally or vertically, all work on the same basic principle. A quick yank on a coiled nylon cord turns a pulley in the starter. The pulley turns a starting gear or a clutch, which spins the crankshaft to begin the internal-combustion cycle that makes the engine run. When the rope is released, a recoil spring causes the pulley to spin in the opposite direction and the rope rewinds automatically.

The system is simplicity itself, and although some small engines are designed with electric starters for convenience, the rope starter is favored for many lawn mowers, chain saws and other small tools. Not only does it work efficiently and reliably, but when it does go on the blink it is easy to repair with nothing more than a wrench, a screwdriver, a pair of pliers and a replacement part or two.

The components of a rope starter that are most likely to break are the rope and the recoil spring. A broken rope is obvious enough. If the rope is good but fails to retract, you can assume the spring is broken. Occasionally the brake spring—a bent wire that puts tension on the starting gear—will come loose or break. When this happens the starter rope will pull out and retract, but the flywheel and crankshaft will not turn.

Some starters must be replaced as a unit if anything breaks, but usually you can replace a broken rope or spring without discarding the rest of the mechanism. You will have to disassemble the starter partially or completely, depending on the model; a few of the commoner types are illustrated on these pages. But procedures and configurations vary with manufacturer, tool and model, so check the owner's manual before you take a starter apart. To aid reassembly, keep the parts in order as you remove them, and note whether the rope and the spring wind clockwise or counterclockwise.

Replacement parts are easy to find. Both springs and nylon rope can be purchased from the tool dealer or from a hardware store. In either case, take the old part with you to ensure an exact match. Before installing a new rope, cut it to the same length as the old rope and, to keep the ends from unraveling, hold each end in a match flame until a soft, sticky ball forms; mash the ball with pliers to make it the same diameter as the rope.

When working with a rope starter, take two precautions to ensure your safety. Before you begin dismantling the starter, disconnect the spark-plug wire and secure it so that it cannot accidentally touch the plug. And always be cautious when you work with a spring. As you disassemble a starter, wear protective goggles and position your body, especially your face, as far away from the spring casing as possible before opening it. If you must unwind a new spring in order to install it, take the same precautions before cutting the clips that secure it. If the spring is to be installed while it is still wound, be sure to grip the coils together with pliers before setting it in place.

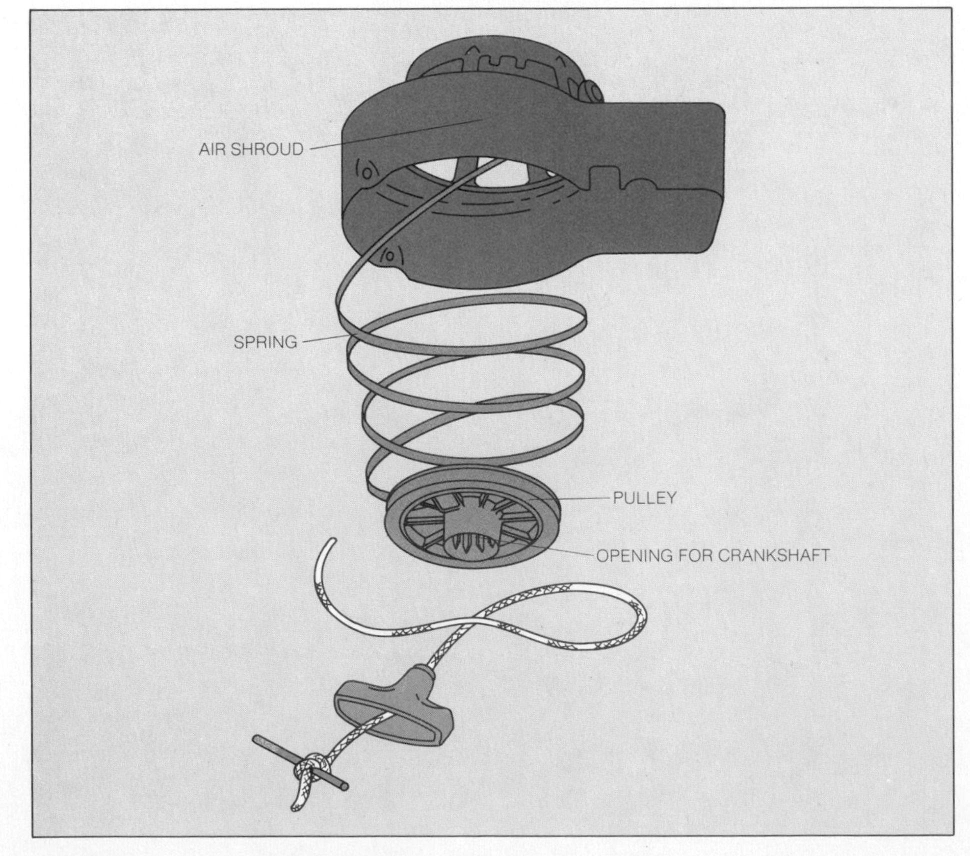

Anatomy of a side-pull starter. In this typical rope-starter configuration, an air shroud conceals a pulley with the starter rope coiled around a groove in its edge. A long, narrow steel spring, with one end hooked to the shroud, lies coiled around a hub on the back of the pulley. When the rope handle is pulled, the rope uncoils and the pulley turns, spinning the crankshaft to start the engine. As the rope uncoils, the spring coils more tightly around its hub on the back of the pulley; when the rope is released, the spring uncoils, forcing the pulley to spin in the opposite direction so that the rope rewinds in its groove.

Replacing the Spring in a Side-pull Starter

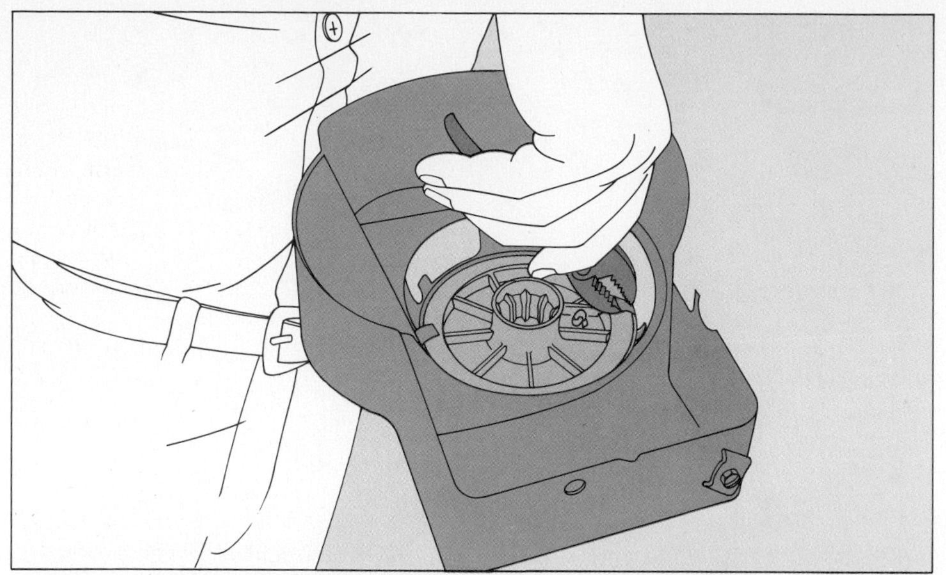

1 Removing the old spring. Take the air shroud off the engine, turn it to expose the rope pulley and use pliers to bend up the metal tabs that hold the pulley in place. If the tabs break, you can use the spares on the shroud to reanchor the pulley later. Slowly lift the pulley out of the shroud, noting the direction in which the spring coils behind it. Unhook the ends of the spring from the pulley and from the shroud, then remove the rope handle and pull the rope from the pulley.

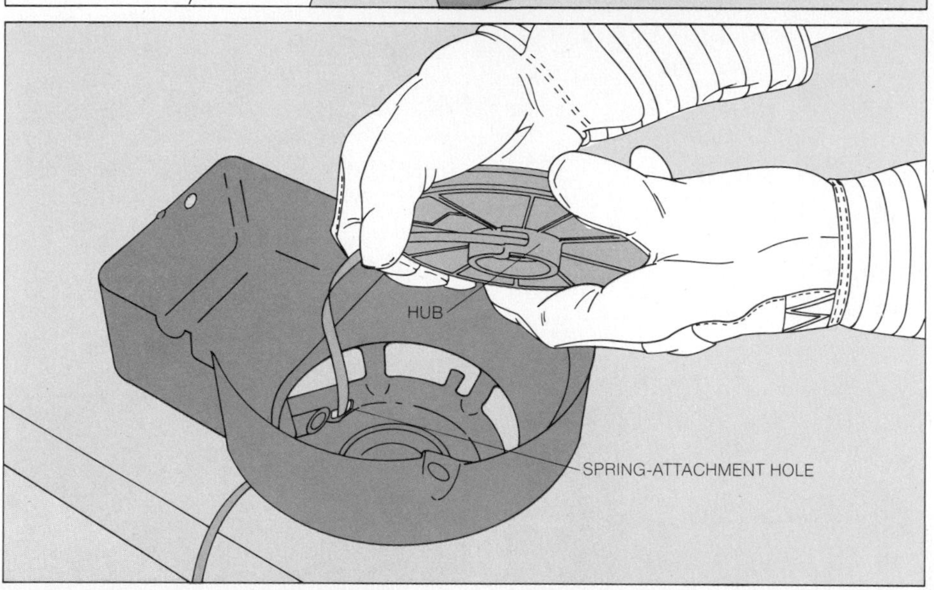

2 Hooking the new spring in place. First straighten the new spring slightly by running your thumb against its inside curve along the entire length of the spring; wear gloves for this, since the spring edges are sharp. Then dab multipurpose grease lightly on the hub of the pulley. Push the end of the spring from the outside to the inside of the shroud through the spring-attachment hole, and hook the spring end to its slot on the pulley hub; make sure the spring is hooked so that it will wind around the hub in the direction noted in Step 1. Set the pulley back in the shroud and press the metal tabs down over its edges, leaving enough of a gap to let the pulley turn freely.

3 Winding the new spring. It is possible to turn the pulley by hand while winding the new spring. The job will be much easier, however, if you cut a rectangular scrap of wood about 4 inches long that fits snugly inside the pulley hub and fit it with a T-shaped handle made from a ¼-inch dowel. Turn the pulley with the wooden T as you feed the spring into the shroud. When no more than 2 inches of the spring protrudes from the shroud, grasp the spring end with pliers for the final few turns, making sure the hooked end of the spring latches securely onto the edge of the spring-attachment hole. Then rewind the starter rope as shown on page 86, top.

Replacing the Rope in a Side-pull Starter

Winding in a new rope. First remove the rope handle and broken rope; you may be able to grasp the ends with pliers and simply yank the parts of the rope out of the pulley. If you cannot because the rope is stuck, lift the pulley out of the shroud *(page 85, Step 1)*, unwind and discard the rope, then put the pulley back in the shroud and recoil the spring *(page 85, Steps 2-3)*.

Prepare the ends of a new rope as described on page 84, then twist the pulley in the shroud until the recoil spring is wound tight. Let the pulley unwind until the rope hole in the pulley lines up with the rope passageway in the shroud, then clamp the pulley in that position with locking-grip pliers. Guide one end of the rope through the pulley hole and the shroud, then pull from the outside to draw the entire length of rope all but a few inches of the way through. Tie off the inside end of the rope, pull it the rest of the way through to the outside and push the knot into the recessed portion of the pulley. Tie the handle to the outer rope end and remove the locking-grip pliers to release the pulley; the spring will uncoil and the rope will wind onto the pulley.

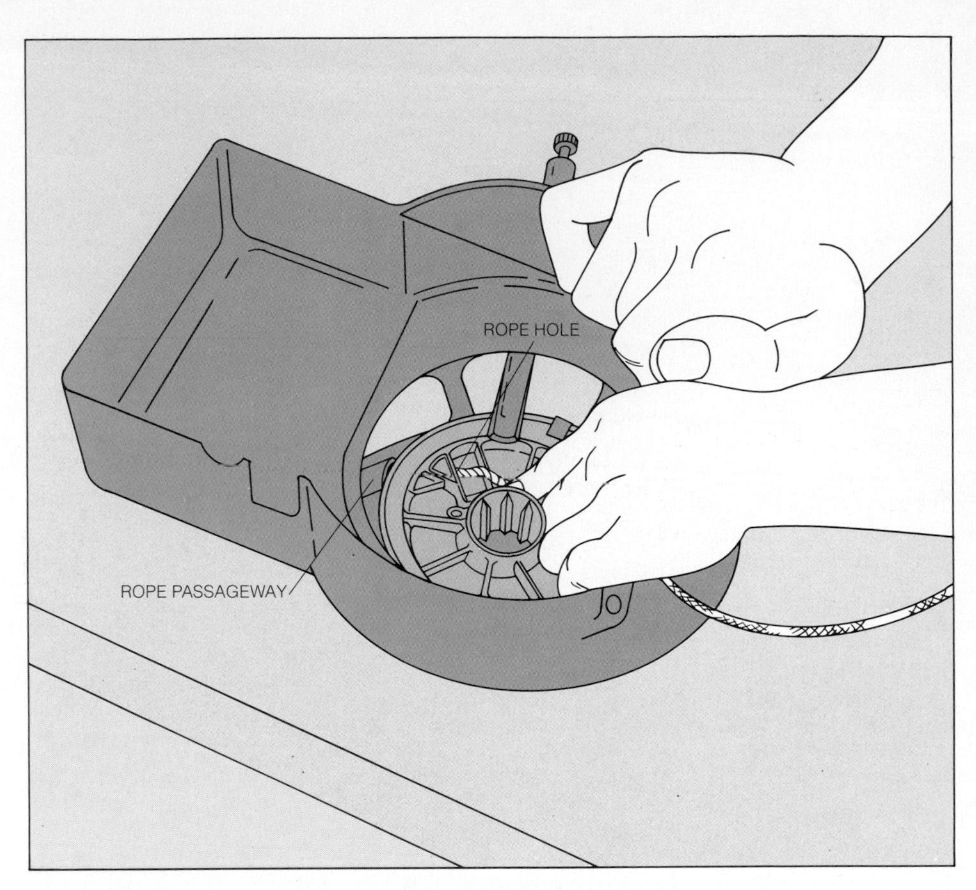

Vertical-pull Starters

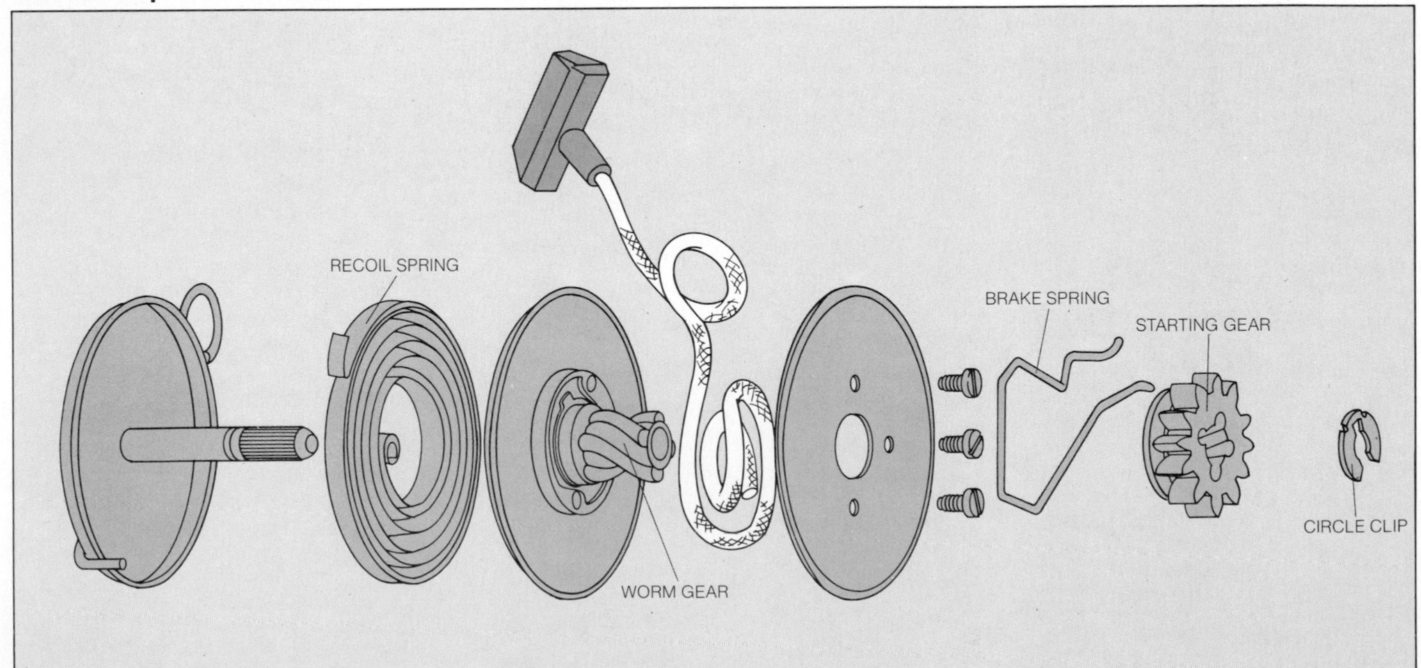

Anatomy of a vertical-pull starter. The recoil spring and the rope on this vertical-pull starter—two layers within a triple-decker sandwich of metal plates—work according to the same principles as the spring and the rope on a side-pull model: When the rope is pulled, the spring winds more tightly, and when the rope is released, the spring uncoils and the rope is wound back up again. A starting gear and a brake spring, which are held in position by a circle clip, are seated on the worm gear at the center of the assembly. When the rope is pulled to make the assembly turn, the starting gear moves on the worm gear to engage the flywheel and turn it, and the engine starts.

Replacing the Spring in a Vertical-pull Starter

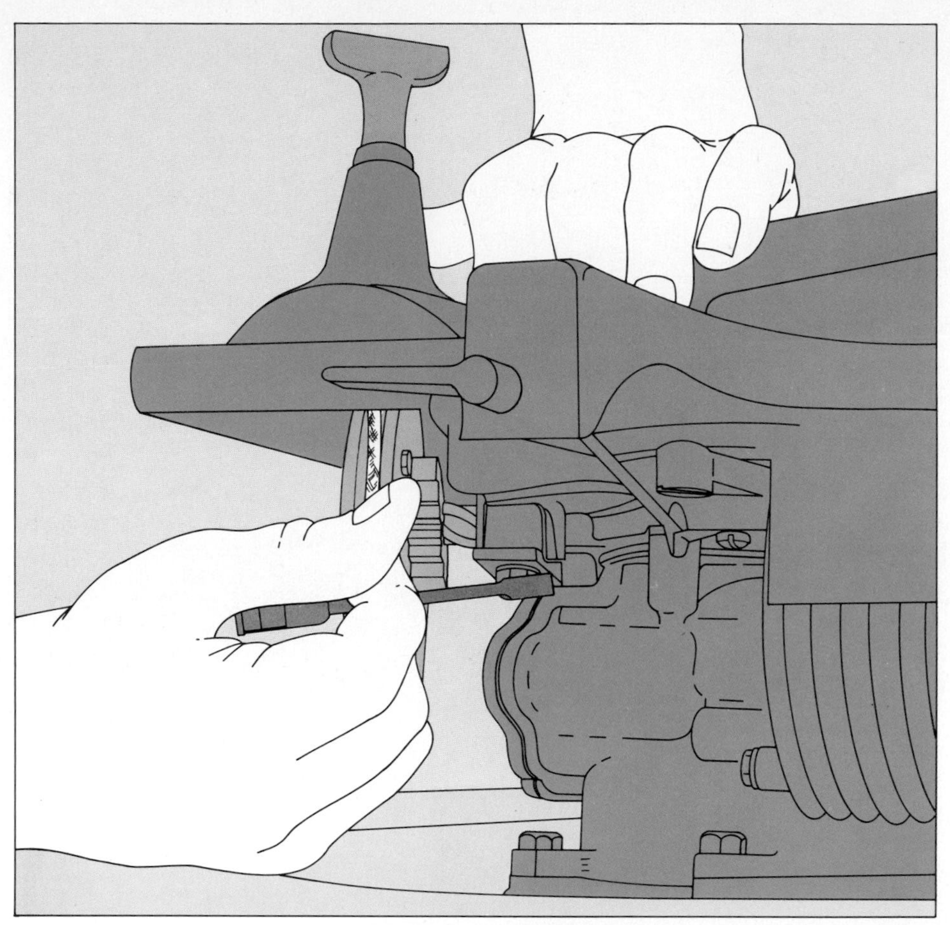

1 Removing the starter. To gain access to the starter on some engines you may have to remove the fuel tank and the carburetor. If you remove the carburetor, stuff a clean rag in the manifold opening to keep out dirt. When the starter is exposed, note the position of the brake spring so that you can replace the starter with the spring facing the same way. Remove the rope handle, then loosen the bolt or bolts that fasten the starter to the engine. Some starters are held in place with allen screws, which must be loosened with a hex wrench. Grip the starter firmly as you remove it to keep the assembly from falling apart.

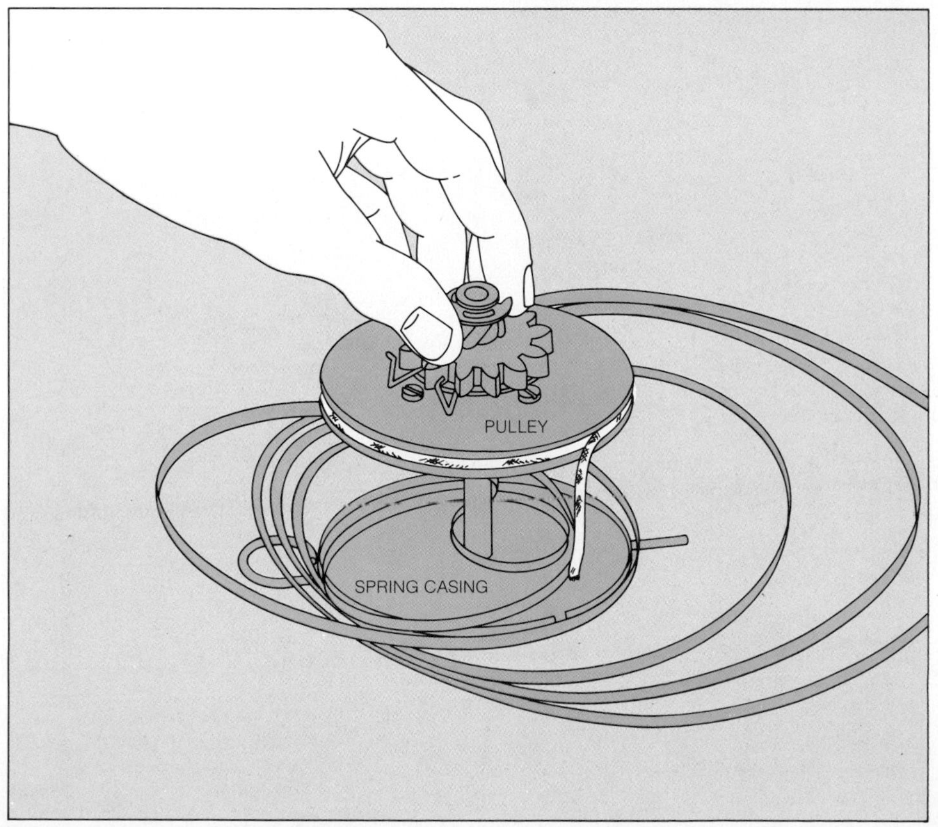

2 Releasing the spring. Continue to hold the parts of the starter together as you lay the assembly, starting gear up, on a flat surface. Stand back from the starter as far as possible, and lift the pulley portion of the starter up and off the central shaft. The spring will pop out of its casing and uncoil; note the direction in which the spring is wound.

3 **Hooking the new spring in place.** First straighten the new spring slightly by running your thumb against its inside curve along the entire length of the spring; wear gloves for this, since the spring edges are sharp. Then flip over the portion of the starter with the worm gear and starting gear; hook the end of the spring to its holder, making sure the spring will coil in the direction noted in Step 2, page 87. Push the shaft of the spring casing into place, making the spring protrude through its slot in the casing.

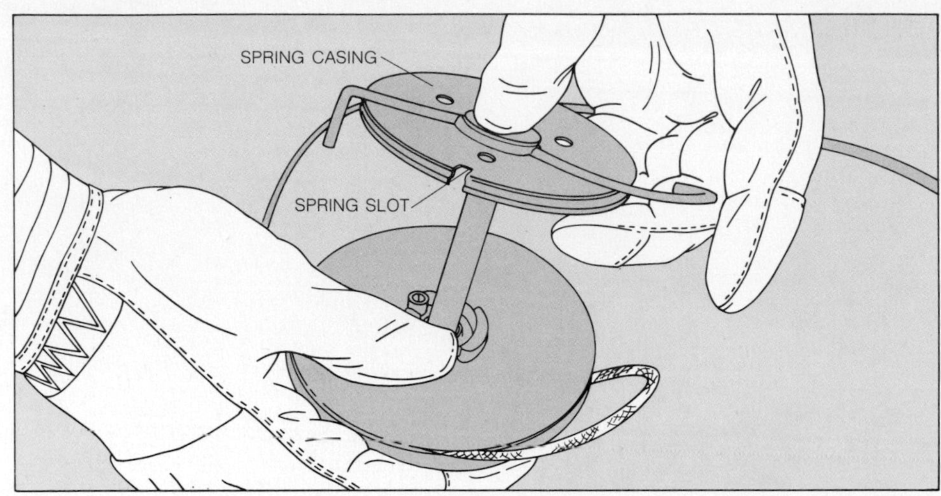

4 **Winding the new spring.** Wind the starter rope around its pulley, and clamp the starter shaft in a vise. Pull the rope toward you; as the pulley turns, the spring will automatically wind into its casing. Rewind the rope as necessary and continue the operation until the spring is fully coiled; hook the end of the spring onto its slot in the casing *(inset)*.

Wind up the starter rope, refasten the starter to the engine and adjust the spring tension by twisting the starter one and a half rotations in the direction in which the rope unwinds. Then tie on the rope handle and replace any other parts removed from the engine.

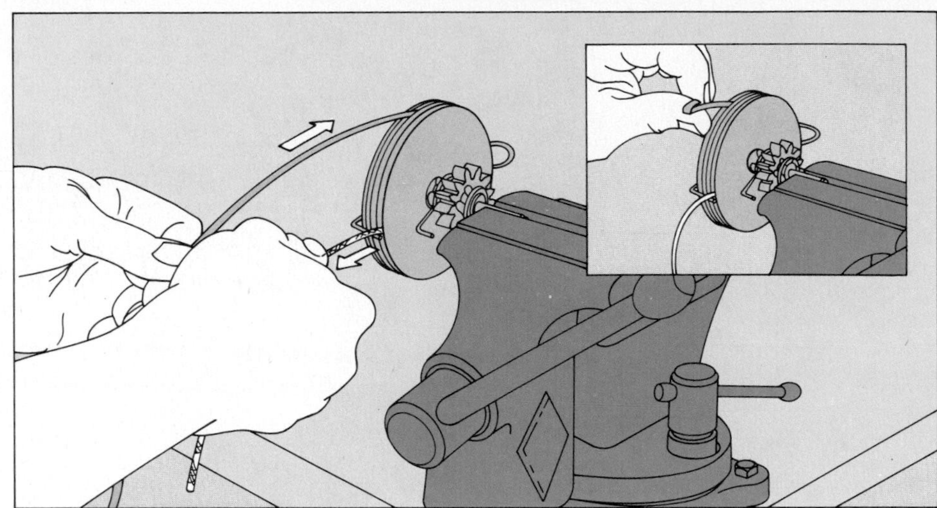

Replacing the Rope in a Vertical-pull Starter

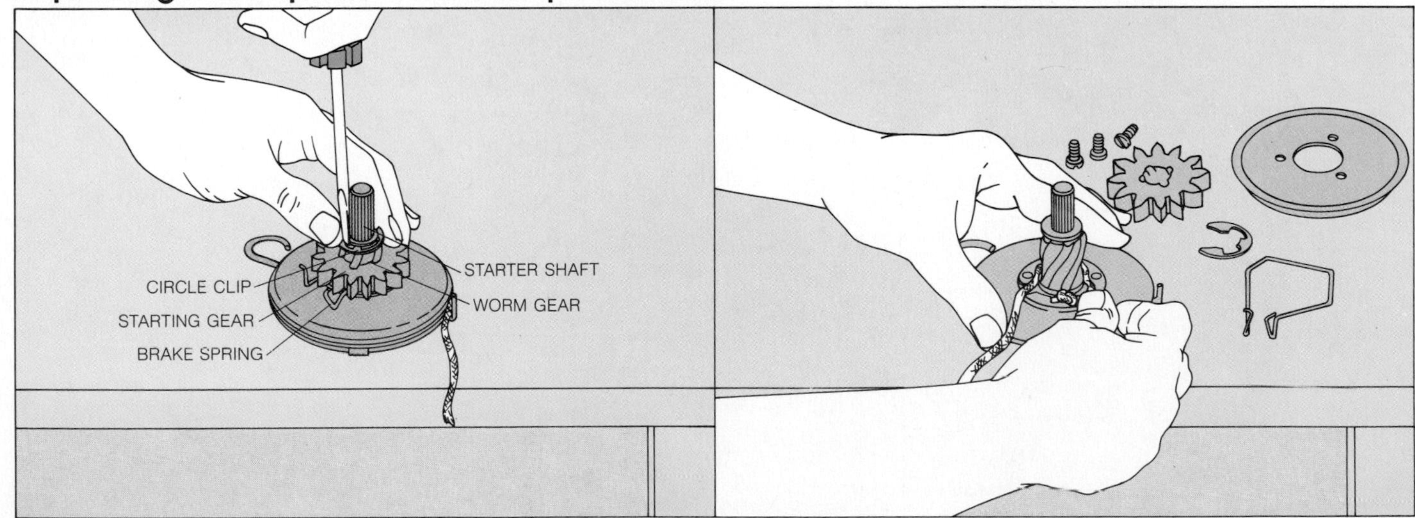

Disassembling the pulley. Remove the starter from the engine *(Step 1, page 87)* and lay it flat, taking care not to lift the pulley off the spring casing. Using the tip of an old screwdriver, pry the circle clip off the starter shaft *(above, left)*, then lift the washer and the starting gear with its brake spring off the worm gear. Finally, remove the screws that anchor the top plate of the rope pulley and lift off the plate. Remove the old rope and push the knotted end of the new rope into its groove around the base of the worm gear *(above, right)*. Put the pulley cover, the starting gear, the washer and the circle clip back into their proper positions, and wind the new rope around the pulley.

Variations on the Basic Designs

A variation on the vertical-pull starter. In the vertical-pull starter shown below, the recoil spring can be replaced with the starter still mounted on the engine. Carefully remove the spring cover and loosen the spring-anchor bolt, noting how the ends of the spring are hooked to the spring anchor and the rope pulley. Then, pinching the coils of the spring together with long-nose pliers, unhook the spring and remove it from the pulley. Grip the coils of the new spring together, cut the clips that are holding it in its coiled position for installation, hook the ends of the spring in place and twist the spring anchor one half turn clockwise to increase the spring tension. Retighten the spring-anchor bolt and replace the cover.

To replace the rope, first remove the spring-anchor bolt and then put back the spring cover to keep the spring from uncoiling. Slip the pulley off its shaft on the mounting bracket and lift the starting gear off the pulley by prying off the circle clip that secures it. Remove the old rope, push the end of the new one into its groove on the pulley and reassemble the starter.

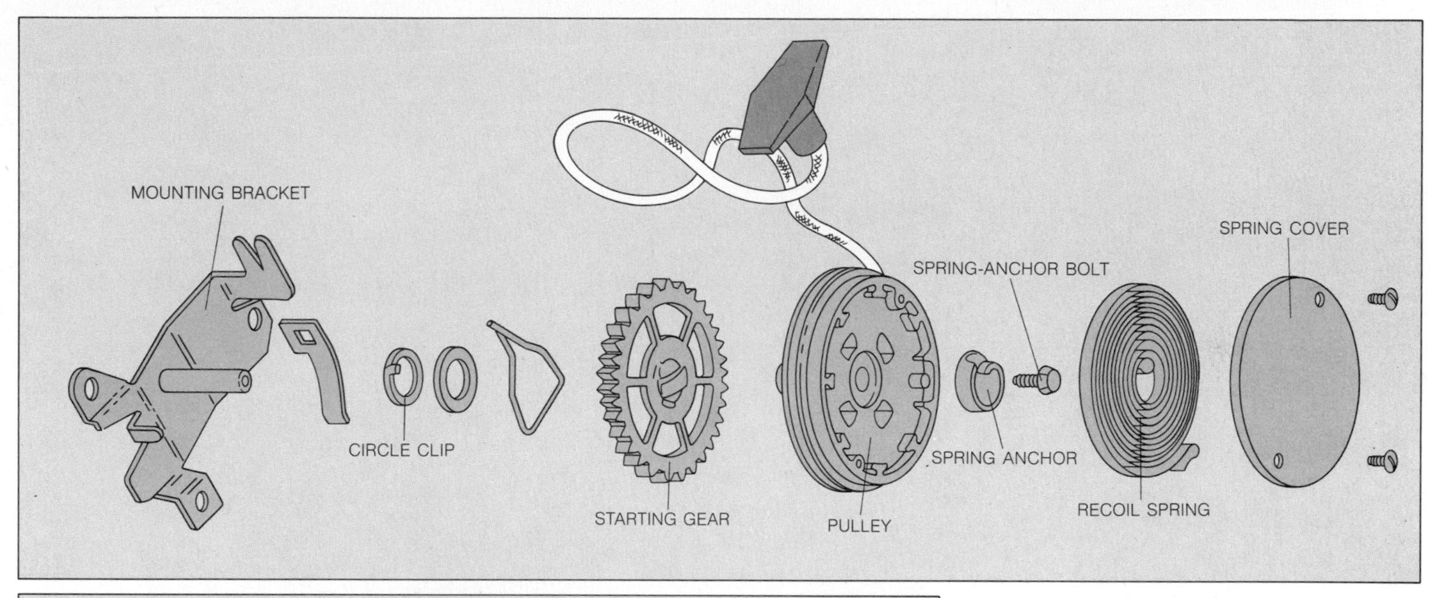

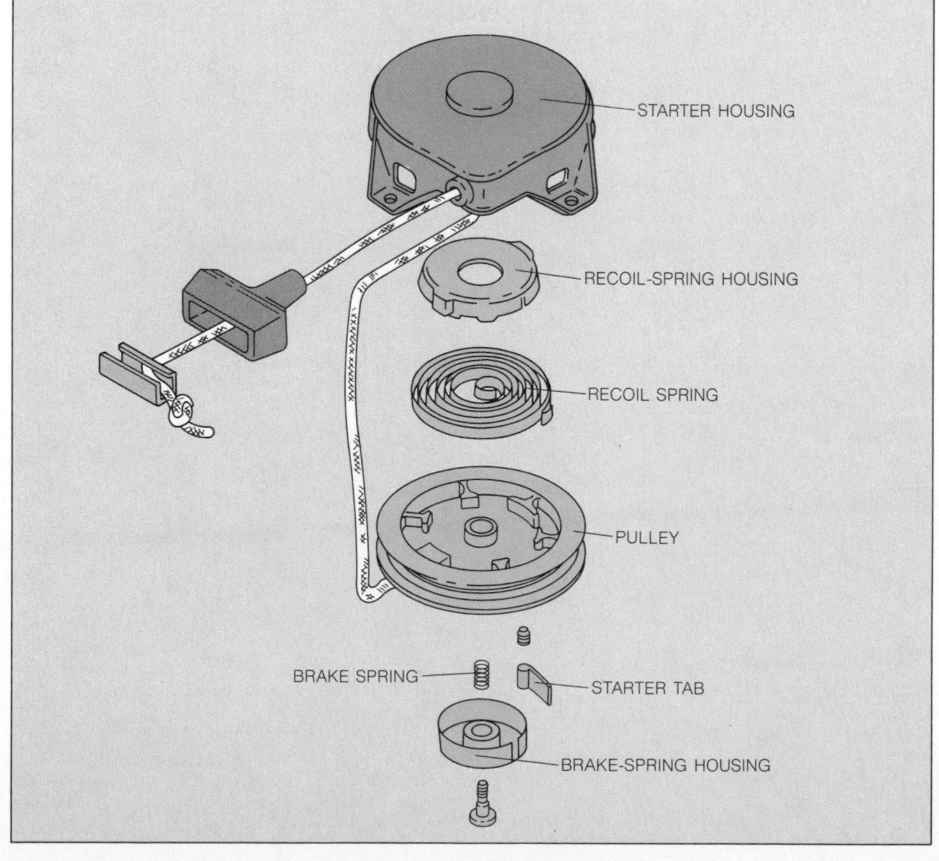

A variation on the side-pull starter. When the rope is pulled on the side-pull starter shown at left, a small, coiled brake spring turns a brake-spring housing, which flips a starter tab. The tab engages a clutch to turn the engine flywheel. On this model, the recoil spring and its housing come as a unit and must be replaced together; the new unit snaps easily into the pulley.

To change a broken rope, disassemble the starter, pull out the old rope, then reassemble the parts. Turn the pulley so that the rope hole in the pulley is lined up with the rope hole in the starter housing; lock the pulley in position with locking-grip pliers as on page 86, top. Thread in the new rope, reattach the handle, then release the pliers; the recoil spring will automatically wind the rope around the pulley.

Electric Starters: Complex but Reliable

Many small engines spring to life effortlessly, without a vigorous tug on a rope, thanks to an electric starter. Such starters are scaled-down versions of automobile starters and are usually energized by a small 12-volt battery. All models operate similarly, turning the engine flywheel by means of a gear or a belt that imparts a slow but strong rotary action until the engine takes over.

Because of their sturdy, uncomplicated construction, electric starters seldom break down. And they need little maintenance. Many have sealed, permanently lubricated bearings and thus need no oiling. When an electric-start engine refuses to turn over, the fault is rarely in the starter itself. Outdoor temperature may be the culprit: In cold weather, batteries put out less energy and oil thickens, making the engine harder to crank. When the temperature drops below freezing, bring the engine and battery inside for a few hours—or rely on the auxiliary rope starter provided with the engine. The fault may also be a weak battery.

Although a few electric starters, particularly those on riding mowers, are equipped with a generator to recharge the battery, most are not. When the battery begins to run down, it must be recharged with separate equipment, which may or may not have been provided with the engine. If your battery needs recharging, do it yourself or take it to a lawnmower or motorcycle service shop, not to an automobile service station. The powerful chargers at many service stations are adjusted for large automotive batteries and can easily ruin a small battery with an accidental overcharge.

If a starting problem is not due to a bad battery, look next at the starter motor. On many small engines, you can clearly see the starter's pinion gear—a large metal or plastic gear mounted on the exposed motor shaft—which cranks the flywheel *(opposite)*. On those models in which the pinion gear is concealed, the gear makes an unmistakable whirring noise when the motor is running. If this gear does not turn, or turns slowly or erratically, the starter motor probably needs a tune-up *(pages 92-93)*.

Begin tuning the starter by removing the end cap, where power enters the starter motor. Inside are a pair of brushes, actually small chunks of carbon that are kept in constant contact by springs with the commutator—a ring of copper bars at the rear of the spinning armature. Eventually these brushes become so worn by friction that they do not make firm enough contact with the commutator to pass current to the armature's copper windings. They and the springs that push them against the commutator must be adjusted or replaced.

A dirty commutator also can prevent electricity from flowing properly. If the commutator protrudes through the open end of the starter housing—or if it is accessible through large openings in the housing—it can be cleaned in place with a folded piece of fine sandpaper. Otherwise, the armature must be removed from the housing before the commutator can be cleaned.

Replacing the armature in an electric starter is much trickier than taking it out. The commutator must be fitted into the end cap, a feat requiring patience and manual dexterity: To push the delicate brushes out of the way, you may need a helper. Sometimes two pairs of hands are needed to fit the combined armature and end cap back into the starter housing. Magnets inside the housing pull at the armature so powerfully that positioning it properly often takes several tries.

If the starter motor turns but does not move the engine flywheel, inspect the teeth of the pinion gear and the flywheel ring gear to see if they mesh. If they do not, the pinion gear is probably worn or damaged; the sturdier and less vulnerable flywheel ring gear rarely, if ever, needs attention. When the gears do not mesh, replace the entire gear assembly by removing the lock nut at the end of the starter shaft and slipping the old parts off and the new ones on.

A few starters have a belt and pulley instead of a pinion gear. If the belt is slipping or worn, tighten it, or remove and replace it, by sliding the starter motor toward or away from the belt after loosening the mounting bolts.

The parts of a starter. A typical small-engine starter motor—mounted here on a lawn mower and wired to a battery fixed to the mower handle—is placed so that its pinion gear spins the ring gear on the flywheel. The end cap of the motor housing holds the brushes, small blocks of carbon that make contact with the commutator to deliver current to the windings. When the current spins the armature inside the housing, the pinion gear turns as shown *(bottom)*.

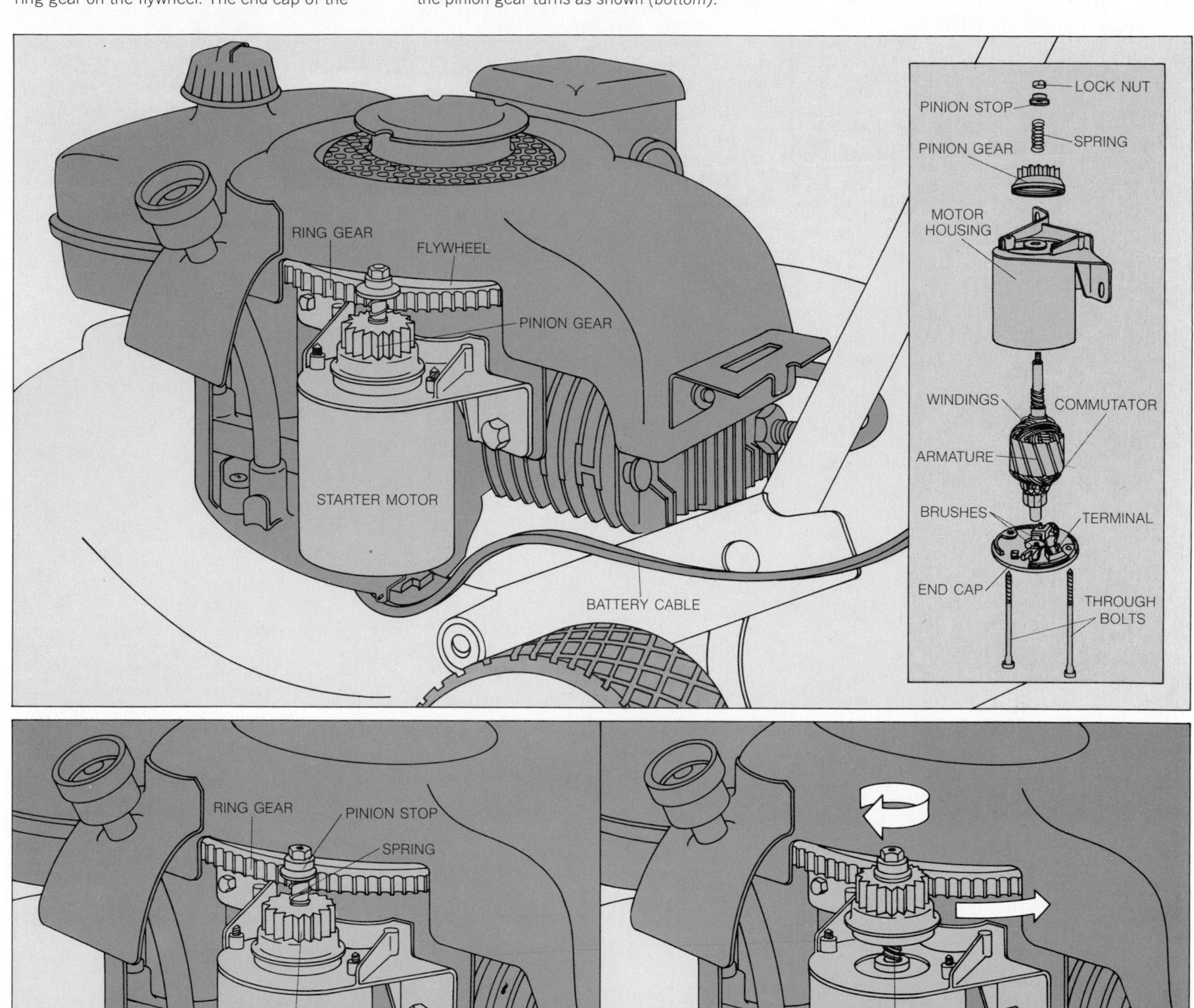

How a pinion gear turns a flywheel. As the armature begins to rotate, the pinion gear momentarily stays motionless because of inertia and a spring that presses on it *(above, left)*. But in less than a second the motor reaches top speed, propelling the gear upward on spiral grooves to mesh with the ring gear of the flywheel *(above, right)*. The pinion gear turns the flywheel, cranking the engine until ignition begins. As the engine starts, the flywheel begins to spin faster than the starter motor; the pinion gear is forced back down to its original position.

Rehabilitating the Starter Motor

1 Examining brushes and springs. After disconnecting the battery cable, unbolt the starter housing. Remove the long through bolts that secure the end cap *(below, left)*, and take it off. Press the brushes back into their housings with your finger, then let them pop out *(below, right)*; if the brushes do not pop out, adjust or replace the springs. Finally, examine the brushes for wear. If the brushes are shorter than they are wide, they should be replaced.

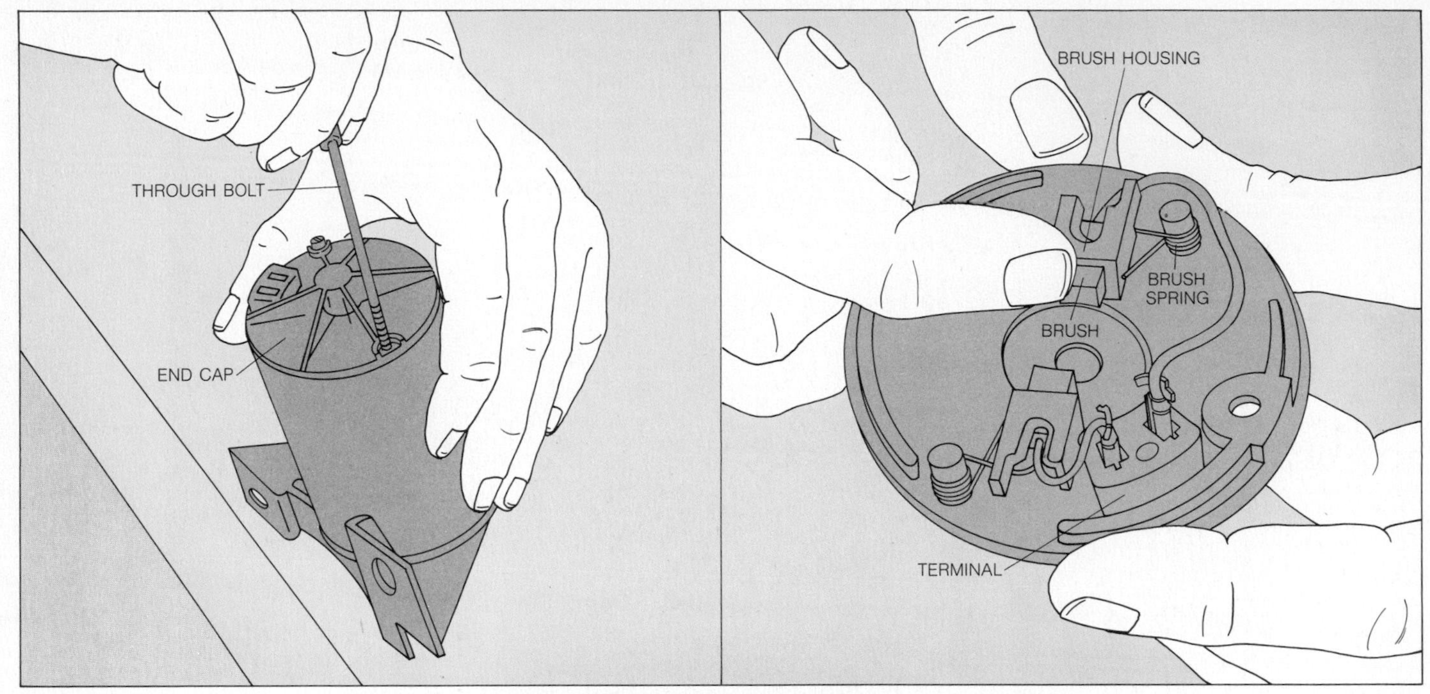

2 Replacing brushes and springs. Apply pressure to the exposed arms of the springs, using a screwdriver to force them back out of the brush housings. Pull the brushes out of their slots and detach the wires from the terminal. Place the new brushes and their attached wires into the slots; install the new springs on the spring posts, using the screwdriver to position one end of each spring against the brush and the other end against the outside of the brush housing. Test the spring tension as in Step 1. Connect the free ends of the wires to the terminal.

On some starters, the wires are held on the terminal by riveted connectors instead of clips or screws. To free the wires, you must drill out the connectors; replace the connectors or solder the replacement wires directly to the terminal.

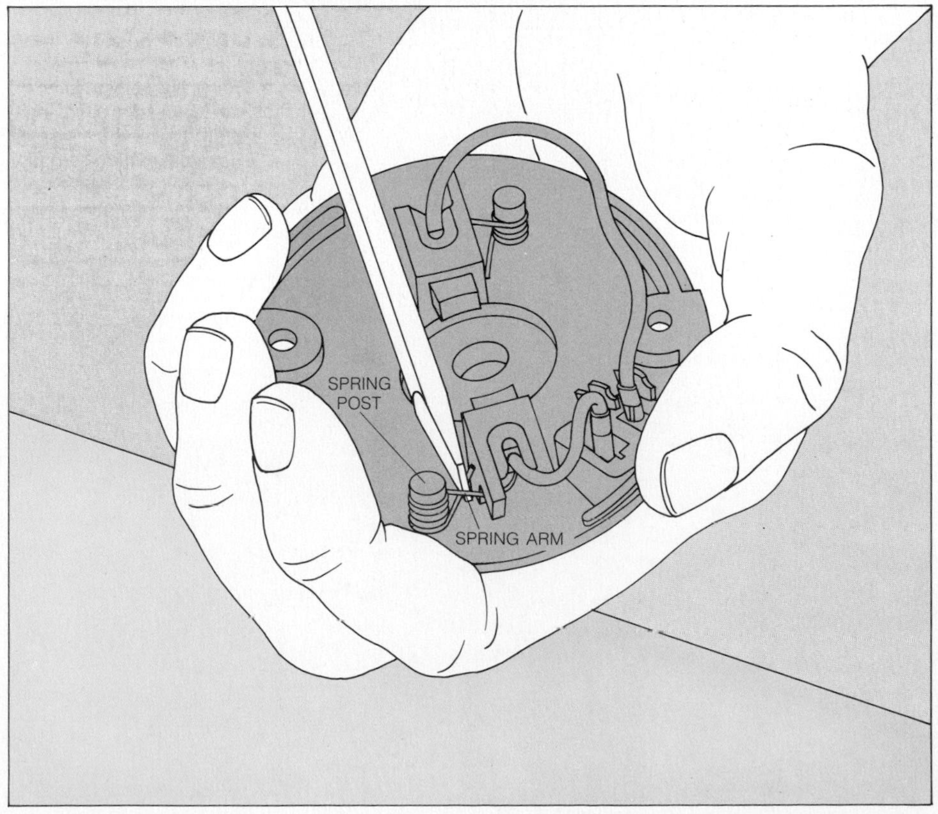

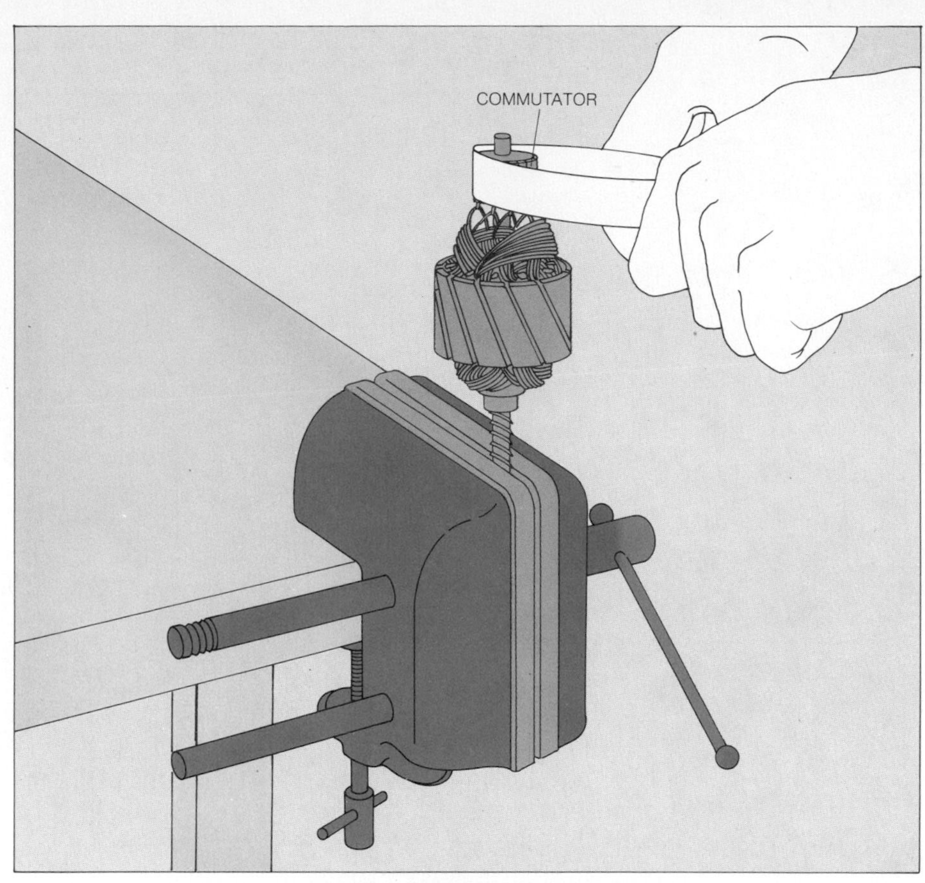

3 Cleaning the commutator. Detach the pinion-gear assembly from the armature shaft by removing the lock nut, then slide the armature out of the starter housing. Clamp the armature in a vise lined with wood to prevent damage to the spiraled grooves on the shaft. Cut a piece of fine (200-grit) sandpaper about 1 inch wide and 6 inches long, and gently work it back and forth—like a shoe-polishing rag—over the commutator, until the copper surface of the commutator gleams. Wipe off any grit or dust that accumulates on the armature, and use a wooden toothpick to clean out the commutator grooves.

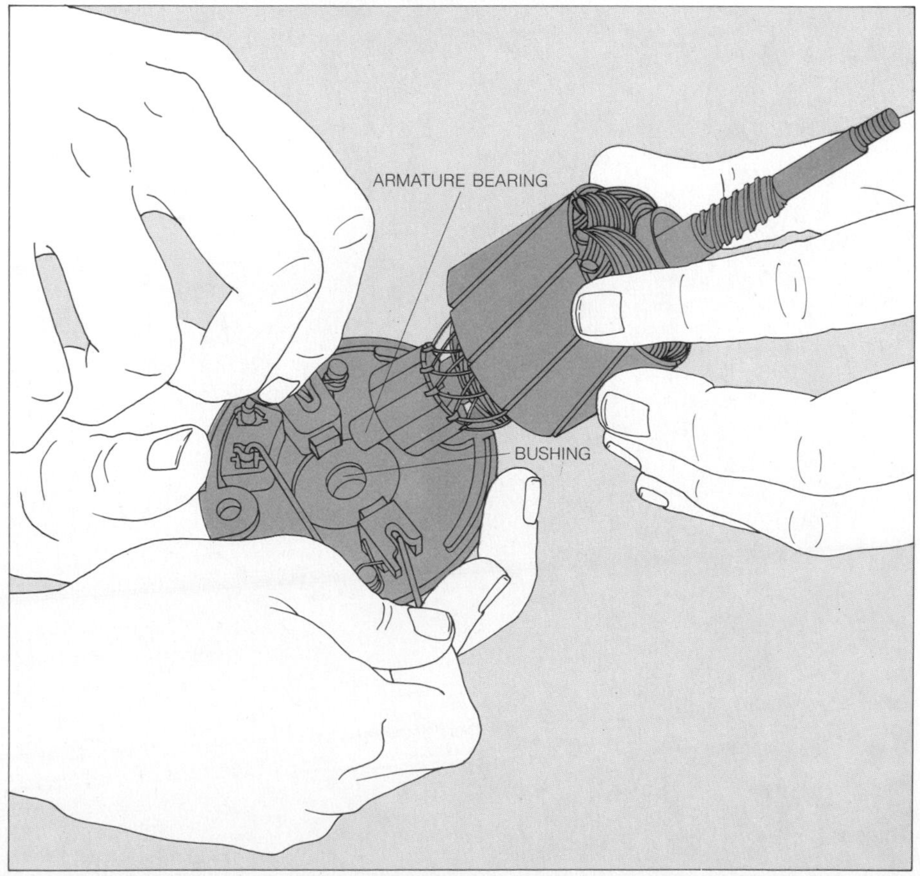

4 Reassembling the starter. With a helper pulling the brushes back into their housings by their wires—or pushing the brushes in with a finger—fit the armature bearing snugly into the recessed bushing at the center of the end cap. Release the brushes and very carefully slide the armature and attached end cap into the starter housing. Twist the end cap to align the holes for the through bolts. Attach the pinion-gear assembly, and slide the pinion gear up the shaft to check the return action of the spring before mounting and connecting the starter.

4 Putting an Engine to Work

The intricacies of a drive system. In the belt-drive transmission system of a power tiller, a set-screw on a pulley attached to the engine crankshaft adjusts the position of the pulley to align precisely with an idler pulley beneath it. A rubber belt, its tension partly determined by this adjustment, is linked to a drive pulley *(not shown)*, which turns the tines of the machine.

A small engine produces power in plenty—from 2 to 20 horsepower, depending on its design and components. But harnessing this power to mow a lawn, till a garden or fell a tree involves more than the engine alone. All machines incorporate a means of transmitting the motion of an engine's driving part, the crankshaft, to one or more driven parts—blades, wheels, tines or the like. Such power-transmission, or drive, systems are made in five basic types, and they are used either singly or in combination on individual machines. In an increasing order of complexity, the five systems are: shaft drive, belt-and-pulley drive, chain-and-sprocket drive, friction drive and gear drive. Each of the systems has its own distinctive problems of service and repair.

Shaft drive, the simplest, is typified by a rotary lawn mower, in which blades are bolted directly to the crankshaft and are turned by the shaft at speeds of 3,000 rpm or more. In such a system, wear and tear on the engine is largely determined by the condition of the blades, which must be kept clean and sharp.

Belts and pulleys are used in power tillers *(left)* and as a second drive system in self-propelled lawn mowers, in which the engine drives the blades directly but turns the wheels with a belt drive. Modern transmission belts are quiet, flexible and strong; the best of them have steel or fiberglass cords to resist stretching. Nevertheless, they require regular inspection for cracks; the pulleys on which they run must be tightened periodically; and you will occasionally have to replace a belt or pulley entirely.

A chain-drive system, used to turn the wheels in some self-propelled lawn mowers and in heavier machines such as riding mowers and lawn tractors, is stronger and more efficient than a belt drive but requires more maintenance. The chains can transfer 98 per cent of the energy of the engine to the driven part, but only if the sprockets are perfectly aligned, the chain links are regularly lubricated and the chain tension is properly adjusted.

A friction system, used to turn the rear wheels on some riding mowers *(page 116)*, consists of a disk driven by a large plate mounted directly on the crankshaft; chains and sprockets transmit the motion of the disk to the wheels. Here, the recurrent problems are adjusting the relative positions of the disk and the plate and maintaining friction between the two. The most complex drive system, a gear system, is usually found on relatively expensive and sophisticated lawn-mowing equipment. The gearbox, a simplified version of the corresponding assembly on an automobile, is usually trouble-free, but you may have to open it occasionally to replace a worn or damaged gear *(pages 112-113)*.

Keeping a Chain Saw Sharp and Running Smoothly

Keeping a chain saw in good working order is an essential task, because a neglected chain saw is as dangerous as it is inefficient. Proper maintenance requires constant vigilance and attention to detail, but none of the work is difficult and much of it is routine. All chain-saw engines and recoil starters are much alike, and procedures for their maintenance and repair are described in Chapters 2 and 3. The parts that make a chain saw unique are the ones that need the most attention: the flexible, multitoothed chain that whirls at high speed around the tempered-steel guide bar; the clutch assembly that starts and stops the spinning; and the oil pump, which keeps the chain lubricated and running smoothly.

The cutting chain generally needs more maintenance than any other part of the saw. Sharpening is a major part of the task; dull cutter links put extra strain on both the operator and the engine. The chain must also be lubricated and adjusted to the correct tension at all times. A chain spinning without enough oil will damage both the guide bar and the drive sprocket. If the chain is too tight, the engine will not accelerate properly and may overheat; if it is too slack, the chain may jump out of its track while the saw is running.

Keeping the chain in good order requires only a few minutes of work with a few special tools each time you use the saw. For sharpening the tiny beveled edges on the chain cutters, you must use a slender, round file—the exact type is specified in the owner's manual or can be recommended by the chain-saw dealer. A special holder for the file has guidelines on top that allow you to file at exactly the right angle. You will also need a single-cut mill file and a depth-gauge tool for filing the depth gauge on each cutter. The file guide and the depth-gauge tool should be purchased with the saw; their sizes and types will be specified in the owner's manual.

Electric chain sharpeners are also available; they vary widely in both price and quality. Such tools may save a few minutes each time you sharpen the chain cutters, but hand filing, for most chain-saw users, produces better results with little extra effort.

When parts of the chain—cutters, drive links or tie straps—are dulled or damaged beyond repair, you can replace them with new parts, available from any dealer or service center. A special notched tool called a chain breaker is used to support the chain while you punch out the old rivets with a pin punch and a ball-peen hammer, and also while you hammer the new parts together.

Lubricating the chain and adjusting the tension are simple tasks. On most saws, the former is done automatically by an oil pump; you need only check the oil level and adjust the flow to ensure proper lubrication. To adjust the flow, follow the procedure on page 105 or, if the oil pump has no dial, use a screwdriver. Checking the chain tension is equally straightforward; if an adjustment is necessary, it can be made quickly with a wrench and a screwdriver.

You must also keep a close watch on the chain guide bar: Cracks, dents or even a moderate build-up of dirt and sawdust may be enough to cause the chain to jam or jump out of its track while the saw is running. You can clean the guide bar and smooth any burrs with a single-cut mill file as described on page 101; but if there are signs of major damage, replace the bar immediately.

Equally in need of constant and careful attention is the clutch assembly, mounted on the saw's crankshaft. The clutch plays an important safety role: It permits the chain to spin only when the operator opens the throttle to start cutting.

The entire assembly works on the principle of centrifugal force, so that only when the engine speed reaches a specified number of rpm's do the clutch shoes, spinning on the crankshaft, push outward enough to engage and turn the clutch case. The clutch case turns the drive sprocket and, consequently, the chain. As soon as the operator releases the throttle the crankshaft slows, the centrifugal force decreases, and the clutch shoes are drawn together by a circular spring. The shoes disengage from the clutch case, and the drive sprocket and the chain stop spinning.

Some clutches do not come apart and must be replaced as a unit, but generally you can remove and disassemble the clutch for servicing. You will need a few special tools. Some, such as hex wrenches to fit the fasteners that hold the outer saw housings in place, may accompany the saw when you buy it; if not, they can be purchased from the chain-saw dealer, as can the clutch wrench designed to grasp and twist the clutch off the crankshaft. You will need a torque wrench *(page 11)* in order to replace the clutch once it has been serviced.

A final indispensable tool for keeping a chain saw in good condition is the owner's manual. It contains specifications for filing angles, the proper track depth for the guide bar, the correct tension for the chain and the recommended torque for the clutch.

The manual will also tell you when to perform routine maintenance tasks, although many requirements are standard. In general, the cutters should be sharpened, the chain tension adjusted and the oil level and flow checked each time you refuel the saw. After several hours of steady use, you should clean and reverse the guide bar and dust the air filter and flywheel *(pages 28-30)*. After about 40 hours of use, the needle bearing in the clutch assembly *(page 104, Step 3)* will probably need grease and the guide bar will need filing, if not replacement. But keep in mind that these are average requirements; always check the owner's manual for the manufacturer's recommendations. When specific problems occur, use the troubleshooting guide *(page 98)* to pinpoint and remedy the trouble.

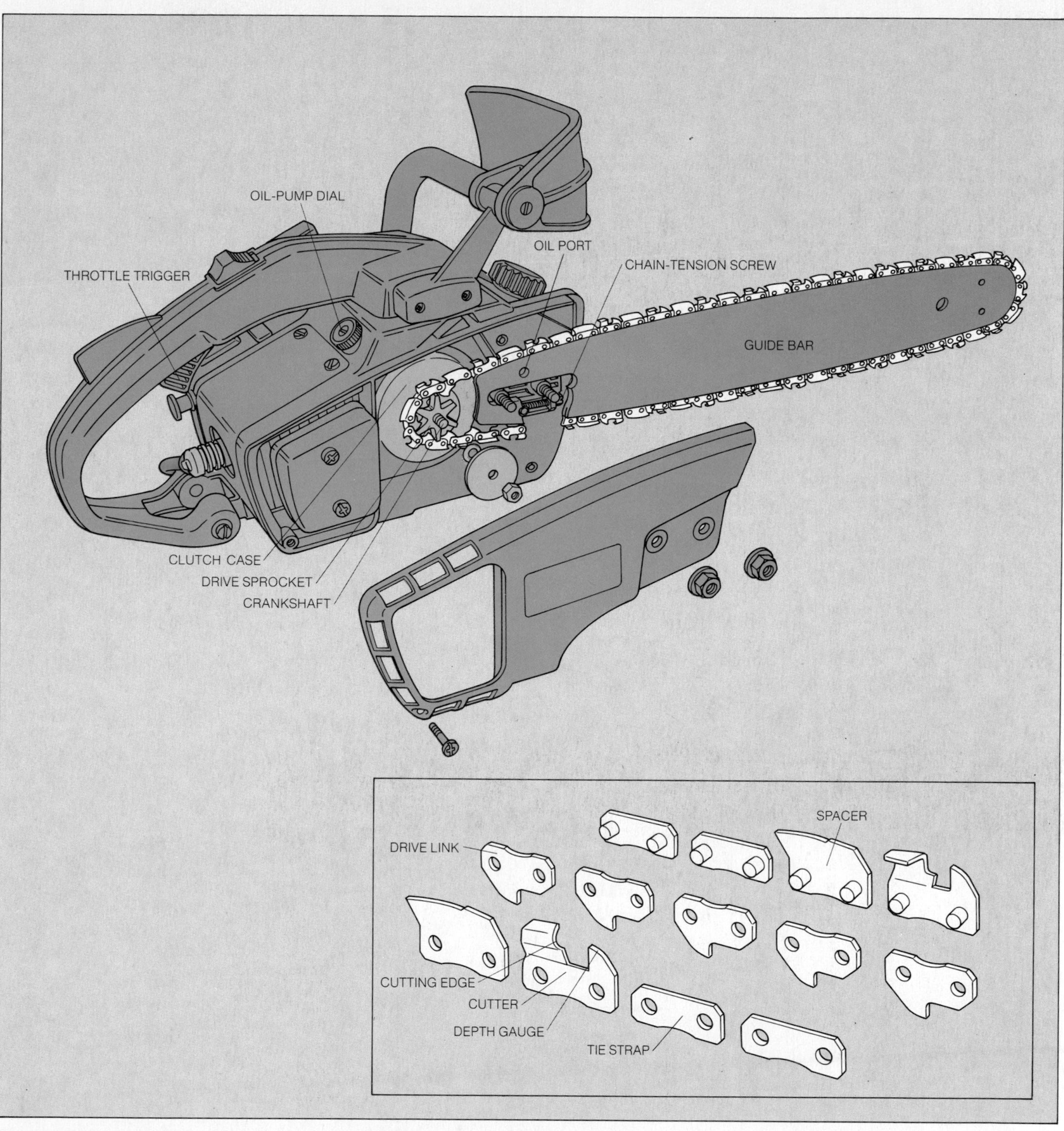

Anatomy of a chain saw. When the throttle trigger is depressed, power from the chain saw's engine spins the crankshaft. Mounted on the crankshaft is a clutch-and-drive-sprocket assembly; once the crankshaft reaches a rotating speed of 2,500 to 3,500 rpm's, this assembly propels the cutting chain in its track around the guide bar. The chain *(inset)* is constructed of three strands of hardened-steel links, riveted together to form one continuous loop. The middle strand consists of a series of drive links whose hook-shaped bottom edges fit into the chain track on the guide bar and engage the teeth on the drive sprocket and the nose sprocket; these links actually move the chain around its course. On either side of the drive links are strands of cutters and tie straps, which maintain proper spacing between cutters. The rounded forward edge of each cutter, called the depth gauge, rides on the bottom of the cut and controls the bite of the beveled cutting edge as it slices through the wood. The tension of the chain on the guide bar is controlled by an adjustment screw at the base of the guide bar.

An oil pump provides a steady flow of oil to lubricate the chain and keep it moving smoothly. The oil reaches the chain through an oil port beside the drive sprocket; the flow is adjusted by means of a dial at the top of the pump.

A Troubleshooting Guide for Chain Saws

Symptom	Probable cause	Remedy
Saw runs erratically	Chain tension too tight	Adjust chain tension *(page 102)*
	Chain binding in guide bar	Inspect and service guide bar *(page 101)*
	Oil pump malfunctioning	Inspect and clean oil pump *(page 105)*
Saw loses power while running	Cutters dull	Sharpen cutters *(opposite)*
	Chain tension adjusted incorrectly	Adjust chain tension *(page 102)*
Chain continues turning while engine is idling	Engine idle speed too high	Adjust idle speed *(page 53)*
	Dirt lodged between clutch and clutch case	Clean clutch assembly *(page 103)*
Chain stops turning (clutch slips) during normal cutting	Chain tension too tight	Adjust chain tension *(page 102)*
	Oil on clutch shoes	Disassemble clutch and clean clutch shoes *(page 103)*
	Too much grease on clutch needle bearing	Disassemble clutch and clean needle bearing *(page 103)*
Drive sprocket excessively worn	Chain tension adjusted incorrectly	Adjust chain tension *(page 102)*
	Clutch needle bearing too dry	Grease clutch needle bearing *(page 103)*
	Guide-bar rails worn or bent	Replace guide bar *(page 100)*
	Cutters dull	Sharpen cutters *(opposite)*
	Not enough oil supplied to chain	Check oil level and add oil if needed *(page 26);* inspect and clean oil pump, adjust oil-pump dial *(page 105)*
Chain insufficiently lubricated	Oil supply low	Add oil *(page 26)*
	Oil port clogged	Clean oil port *(page 105)*
	Oil pump improperly adjusted	Adjust oil-pump dial *(page 105)*
	Fuel line and filter clogged	Clean or replace parts *(page 26)*
	Oil leaking from top of pump	Replace upper neoprene gasket *(page 105)*
	Oil leaking from bottom of pump into engine	Replace lower neoprene gasket *(page 105)*

Using the troubleshooting guide. The left column of the troubleshooting guide above lists symptoms of malfunction that occur commonly in chain saws. The middle column lists possible causes for each symptom. Directly to the right of each cause is the appropriate remedy.

Filing the Cutters and Depth Gauges

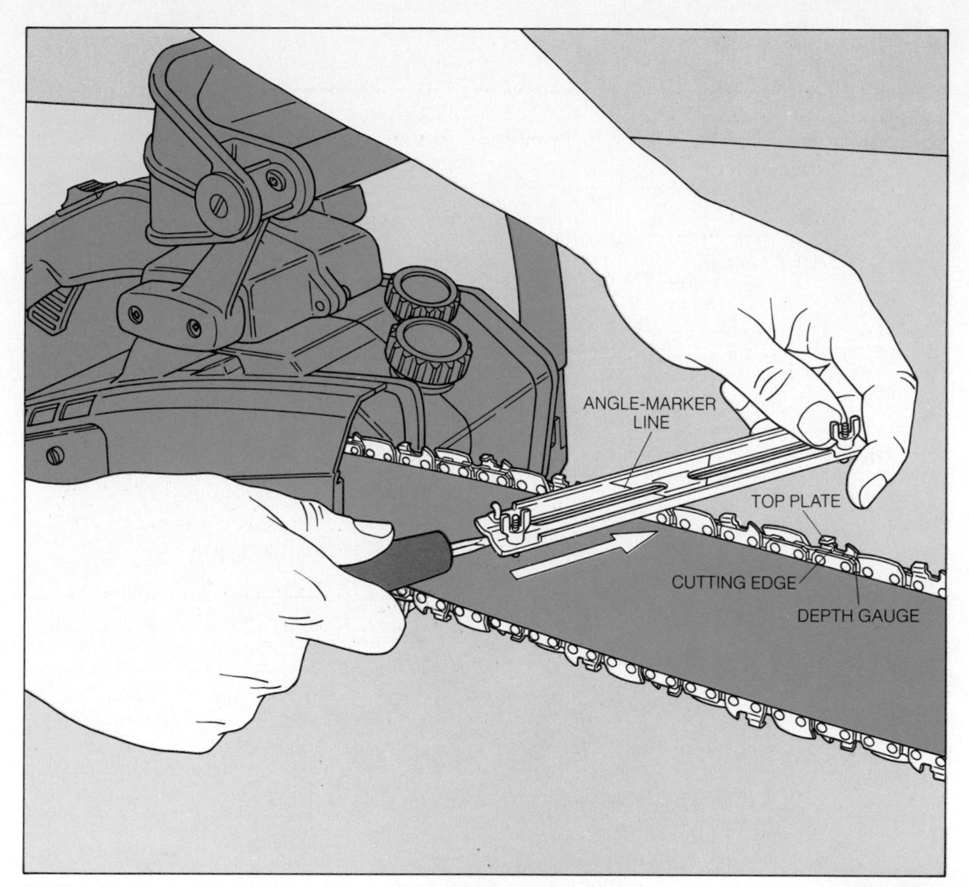

1 Sharpening the cutting edges. Turn the chain-tension screw clockwise until the chain is tight and rigid. Lay the file guide horizontally across the top of the chain so that the file fits against a cutter's curved, beveled cutting edge and the guide is supported by the top plate and the depth gauge. Align the appropriate angle-marker line *(page 96)* with the edge of the guide bar and push steadily forward, applying moderate pressure against the cutting edge. Repeat the same forward stroke two or three times; do not apply pressure to the cutting edge on the return stroke.

File all the cutters that have similarly angled cutting edges—every other cutter on the chain. Then switch the angle of the file guide and go on to file the remaining cutters.

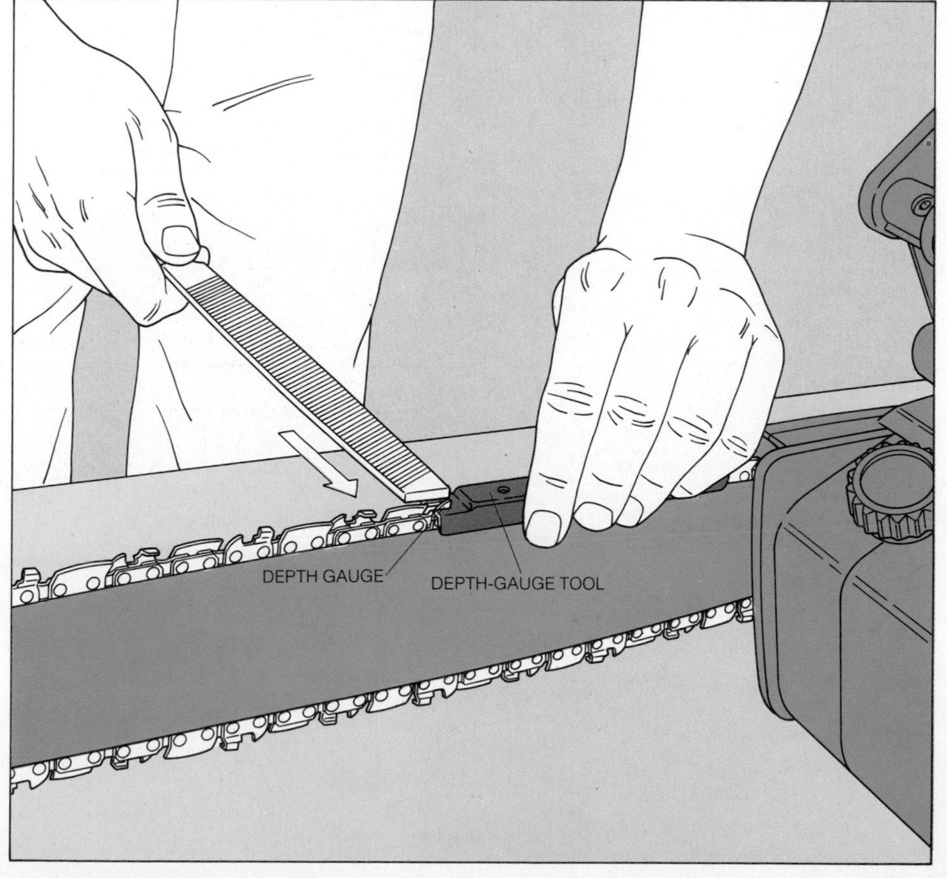

2 Filing the depth gauges. Check the depth gauge on each cutter after you have sharpened its cutting edge. Set the depth-gauge tool over the top of the chain as shown, the back edge of its grooved toe butting against the inside edge of the depth gauge. If the depth gauge protrudes, hold a single-cut mill file horizontal and perpendicular to the chain and push the file lightly across the top of the depth gauge. Lift the file and return it to the starting position; do not pull the file back across the metal edge. Continue filing until the top edge of the depth gauge is level with the surface of the depth-gauge tool, then remove the tool and round off the outer corner of the depth gauge with a few light strokes, again filing only on the forward stroke.

Replacing a Worn or Broken Link

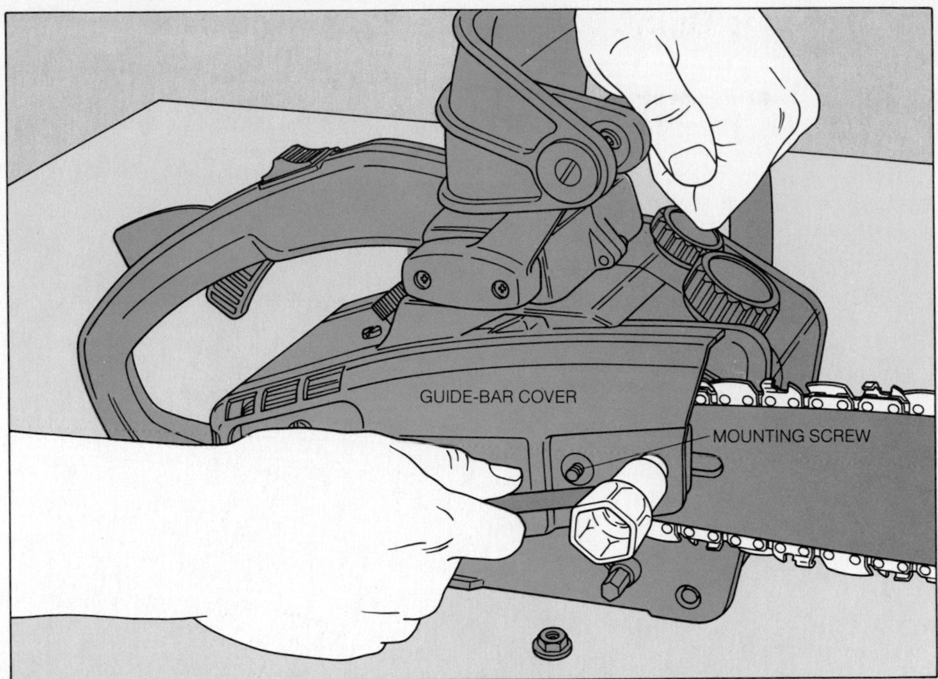

1 **Removing the guide bar and chain.** Turn the chain-tension screw counterclockwise until the chain is completely slack, then loosen and remove the nut or nuts that secure the guide-bar cover and the guide bar on their mounting screws. Pull the cover off the mounting screws; then, as you pull the guide bar off the same screws, lift the chain off the drive sprocket on the saw body. Once the guide bar and chain are freed, lift the chain out of its track.

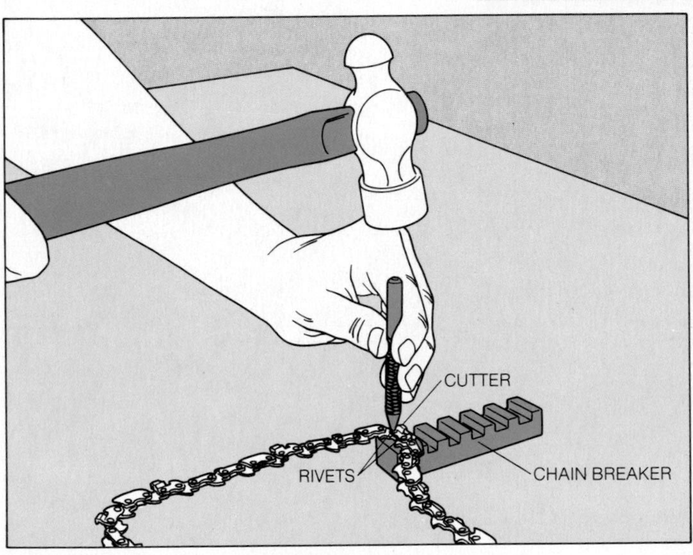

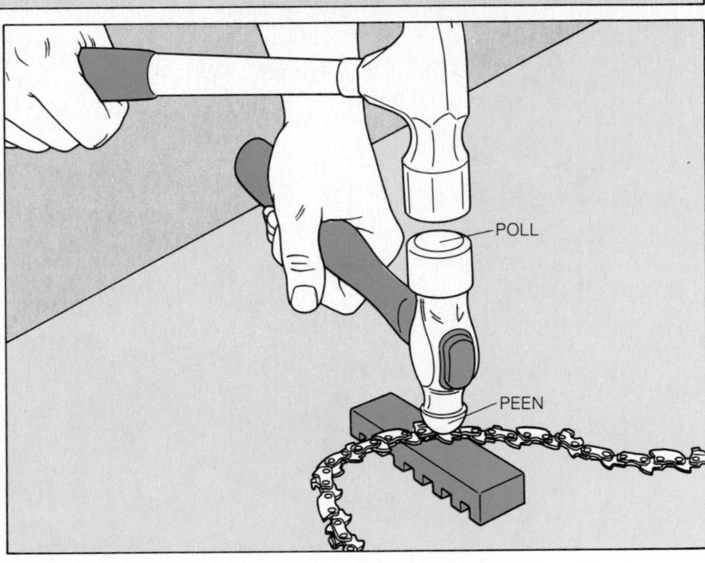

2 **Removing the damaged link assembly.** Set the chain breaker on a flat surface, slotted face up, and position the chain atop the breaker so that the tie strap for the damaged link assembly is underneath and the cutter is on top; the two rivet heads for the link should be cradled in the breaker slot that fits them best (other slots are for different chains). Set the tip of a pin punch on one of the rivets, and tap the punch firmly two or three times; then switch to the other rivet and repeat the process. Alternate between rivets until both are punched out.

Take the damaged link assembly apart and hold the damaged tie strap, drive link or cutter against its replacement to check for variation in size. File down the edges of the replacement part with a single-cut mill file, if necessary, so that the two pieces will match.

3 **Reassembling the chain.** To put the new link assembly together, position the part with rivets, rivet heads down, atop a flat section of the chain breaker and then stack the remaining parts of the link in order on the rivet shanks. Be sure the cutter is on the opposite side of the chain from the two adjacent cutters, and the cutting edge faces in toward the center of the chain. Set the peen of a ball-peen hammer head on the end of a rivet and, wearing goggles, strike the poll with a second hammer until the end of the rivet spreads and the parts of the link assembly are held together but can still pivot freely; hammer the second rivet in the same way. Shape both rivet ends into small knobs by tapping lightly around their edges with the hammer peen.

Before replacing the chain, check the guide bar and service it *(page 101)*.

Servicing the Guide Bar

1 **Cleaning and filing the guide bar.** While the chain is off the guide bar, use a small screwdriver or a stiff wire to poke accumulated dirt and sawdust out of the chain track around the perimeter of the guide bar and out of the oil inlets in the face of the guide bar *(below, left)*. If burrs have formed on the outer edges of the rails, lay the guide bar flat and file the burrs down with a single-cut mill file. Hold the file perpendicular to the edge of the guide bar and move away from the center of the bar *(below, right)*.

Use a steel ruler to measure both the depth of the chain track and the gap between the guide-bar rails all around the guide bar; correct measurements are specified in the owner's manual. If the rail edges are worn down, making the track too shallow at any point, or if the rails are bent too close together or spread too far apart, the guide bar should be replaced. Also replace the guide bar if you find cracks in the rails.

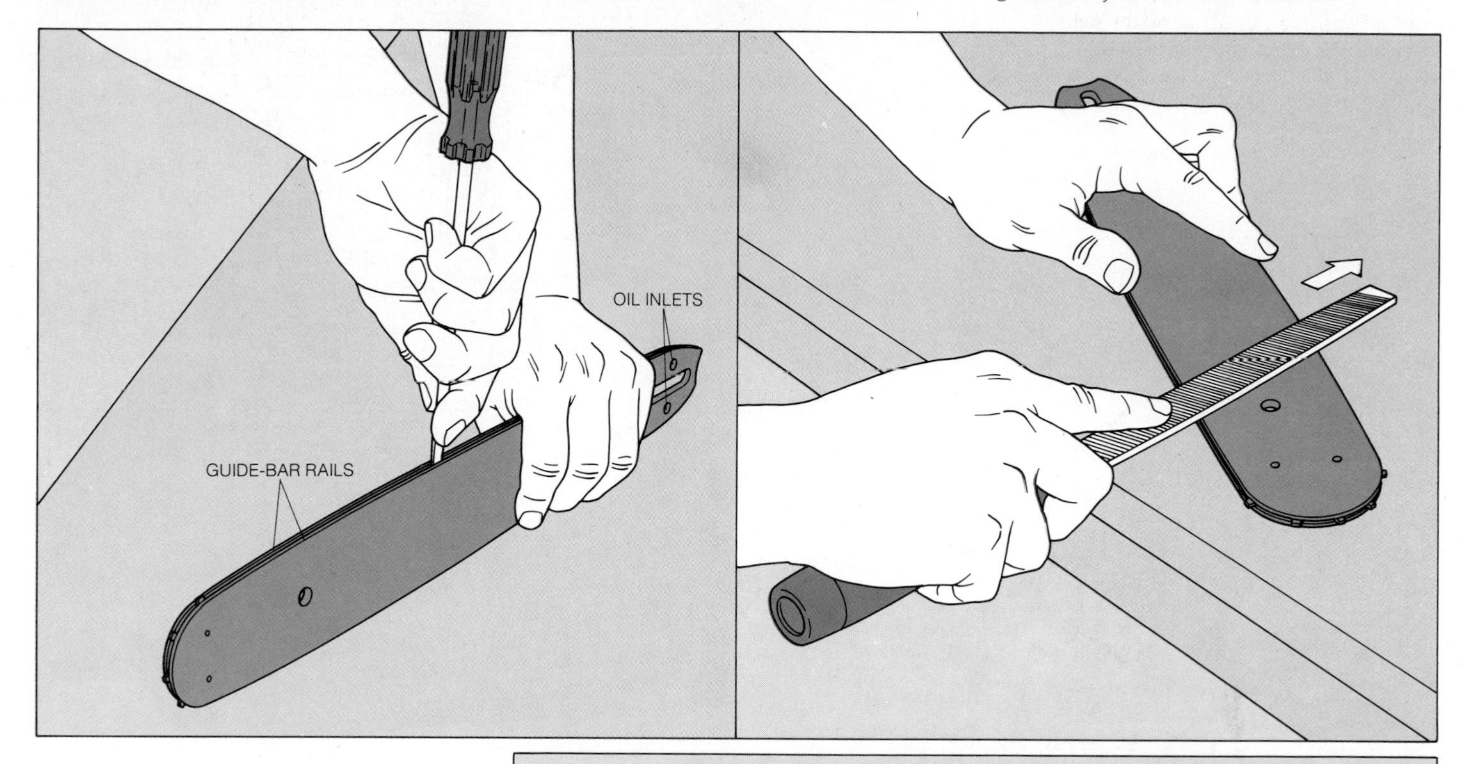

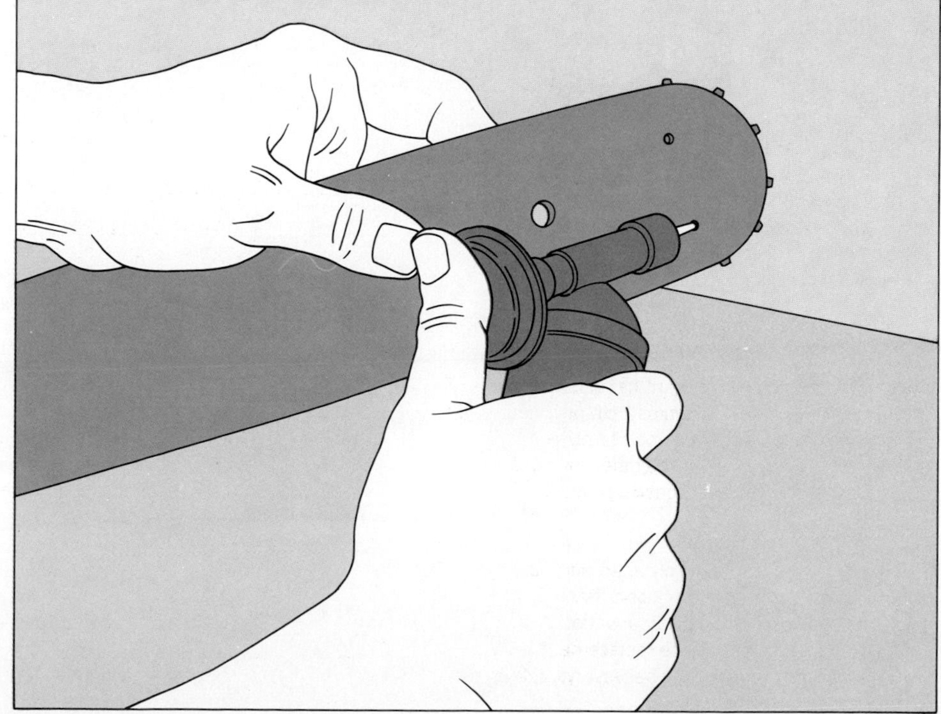

2 **Cleaning and greasing the nose sprocket.** Poke accumulated grease and sawdust out of the inlet near the tip of the guide bar, then push the nozzle of a hand-pump grease gun into the inlet. Turn the nose sprocket as you pump grease into the inlet; continue until grease begins to ooze out around the edges of the bar rails. Flip the guide bar over and repeat the procedure.

3 **Replacing the guide bar and chain.** Attach the guide bar to the saw body by slipping the appropriate openings in the guide bar over the mounting screws and the chain-tension screw pin. Then, wearing gloves to protect your hands, wrap the chain around the drive sprocket and the guide bar. Be sure that the cutting edges on the cutters at the top of the guide bar point toward the tip of the guide bar and that the hook-shaped bottoms of the chain's drive links fit into the chain track and into the teeth of the nose and drive sprockets.

Pull the guide bar away from the drive sprocket so that there is no slack in the chain. Holding the guide bar in that position, give the chain-tension screw a few clockwise twists to immobilize the chain. Still holding the guide bar with one hand, push the guide-bar cover onto the mounting screws, and twist on the nut or nuts that hold it in place. Tighten the nuts with your fingers.

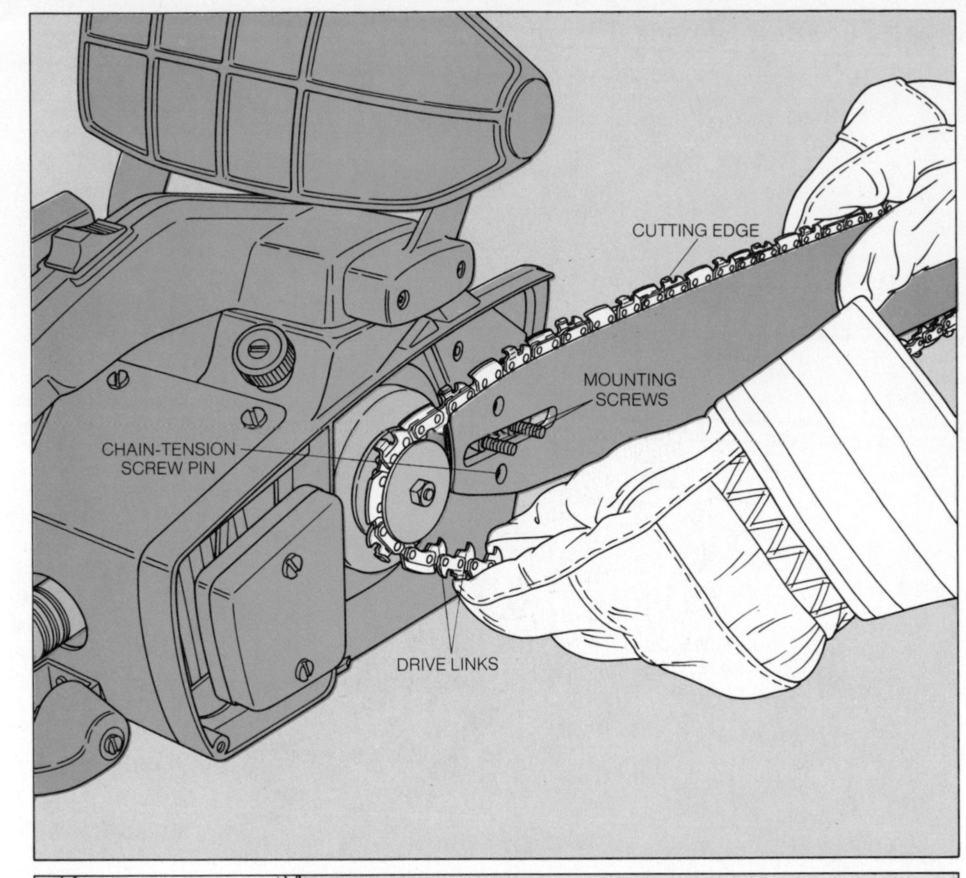

4 **Adjusting the chain tension.** Lift the nose of the guide bar as high as it will go, then turn the chain-tension screw clockwise until the bottoms of the chain's drive links are just even with the edges of the rails along the bottom of the guide bar; tighten the screw an additional half turn. Continue to hold up the nose of the guide bar and tighten the nut on each mounting screw with a wrench. Pull the chain one full circuit around the guide bar; if it does not turn freely, check the position of the drive links in the chain track and on the sprockets, and make any necessary adjustments. Finally, lift the center of the chain at the top of the guide bar as far as it will go; the bottoms of the drive links should be level with the top edges of the rails *(inset)*. Readjust the chain-tension screw if necessary.

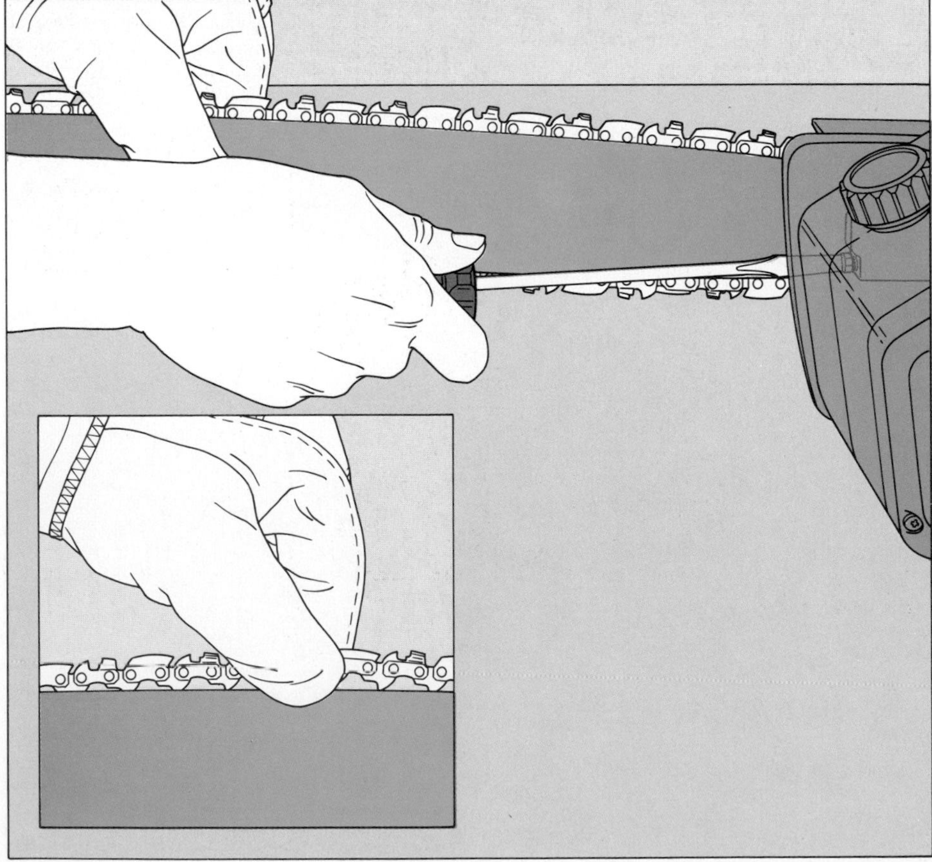

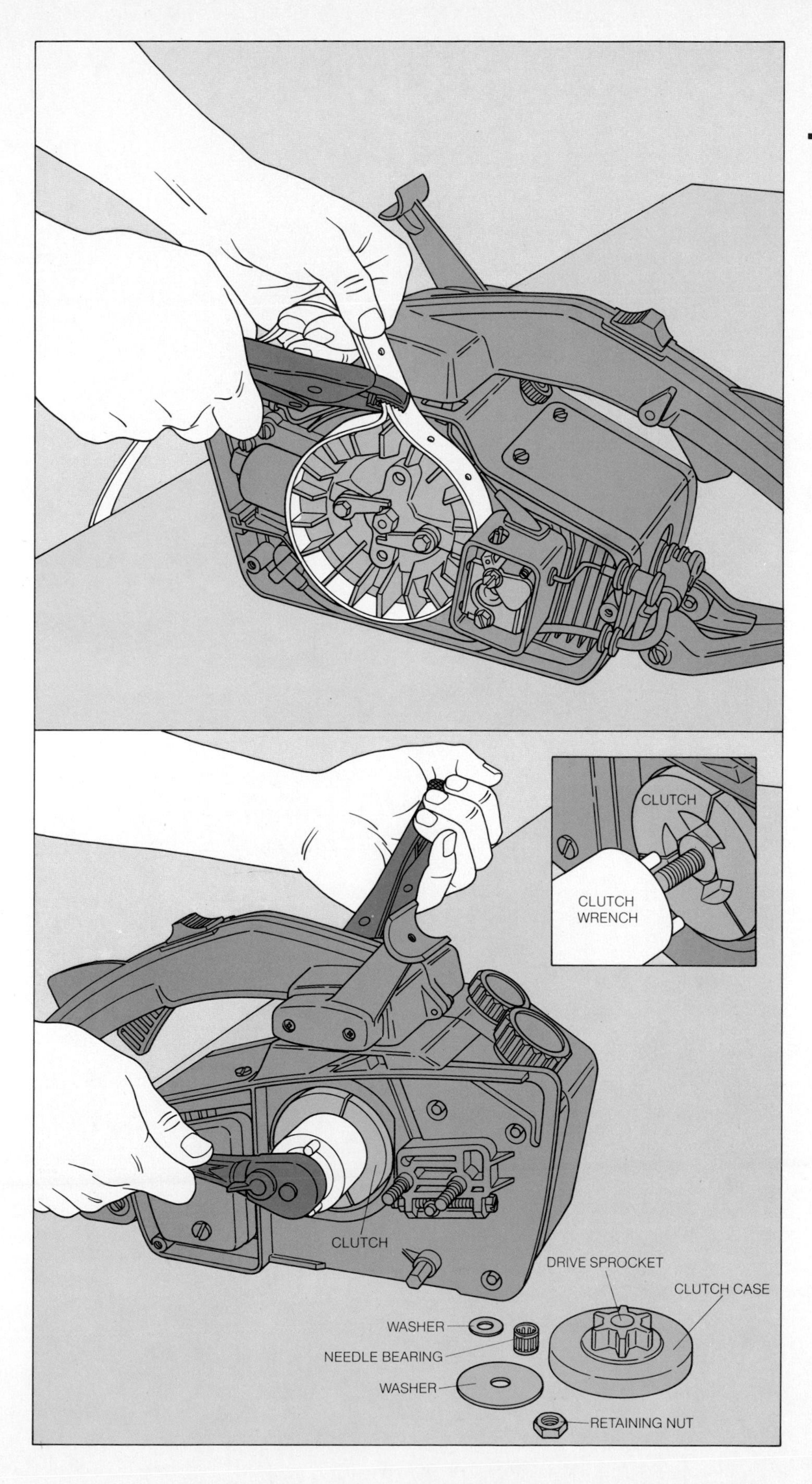

Removing and Repairing the Clutch

1 Removing the clutch assembly. Remove the saw's front hand grip and fan housing to expose the flywheel. On the other side of the saw, take off the guide-bar cover and the guide bar with the chain. Immobilize the flywheel: If it has no central nut that can be held by a socket wrench, secure the wheel by wrapping a leather belt around the fins and holding it tight with locking-grip pliers *(top left)*.

With the flywheel held stationary, loosen the clutch assembly's retaining nut by turning it clockwise; pull the nut and washer, the drive sprocket, and then the clutch case, the needle bearing and the rest of the washers off the crankshaft, noting their order as you set them aside. Insert the pegs on the clutch wrench in the corresponding openings on the clutch *(inset)* and, still holding the flywheel, twist the clutch clockwise *(bottom left)* and pull it off the crankshaft.

2 Checking the clutch for damage. Force the clutch spring out of its groove by pushing it very gently from underneath with the tip of a screwdriver, then pull the clutch shoes off the central hub and inspect them for dirt and damage. Using a screwdriver or a sturdy wire, carefully poke accumulated sawdust out of the grooves in the shoes and hub *(below, left)*, and wipe each part with a rag that has been dampened in kerosene. Inspect the clutch shoes for cracks and worn spots; replace all of the shoes if any one is damaged. Then inspect the needle bearing for missing bearings; replace the part if there is any damage. Finally, inspect the drive sprocket—if the sprocket teeth are worn down, you should replace the part.

To reassemble the clutch, lay the hub down and slide each shoe onto the hub with the groove for the clutch spring facing up *(below, right)*. Hook the ends of the spring together. Stretch it slightly and roll it into the groove with your fingers; do not force the spring with a screwdriver or other metal tool, which could damage the spring coils or scratch the clutch shoes.

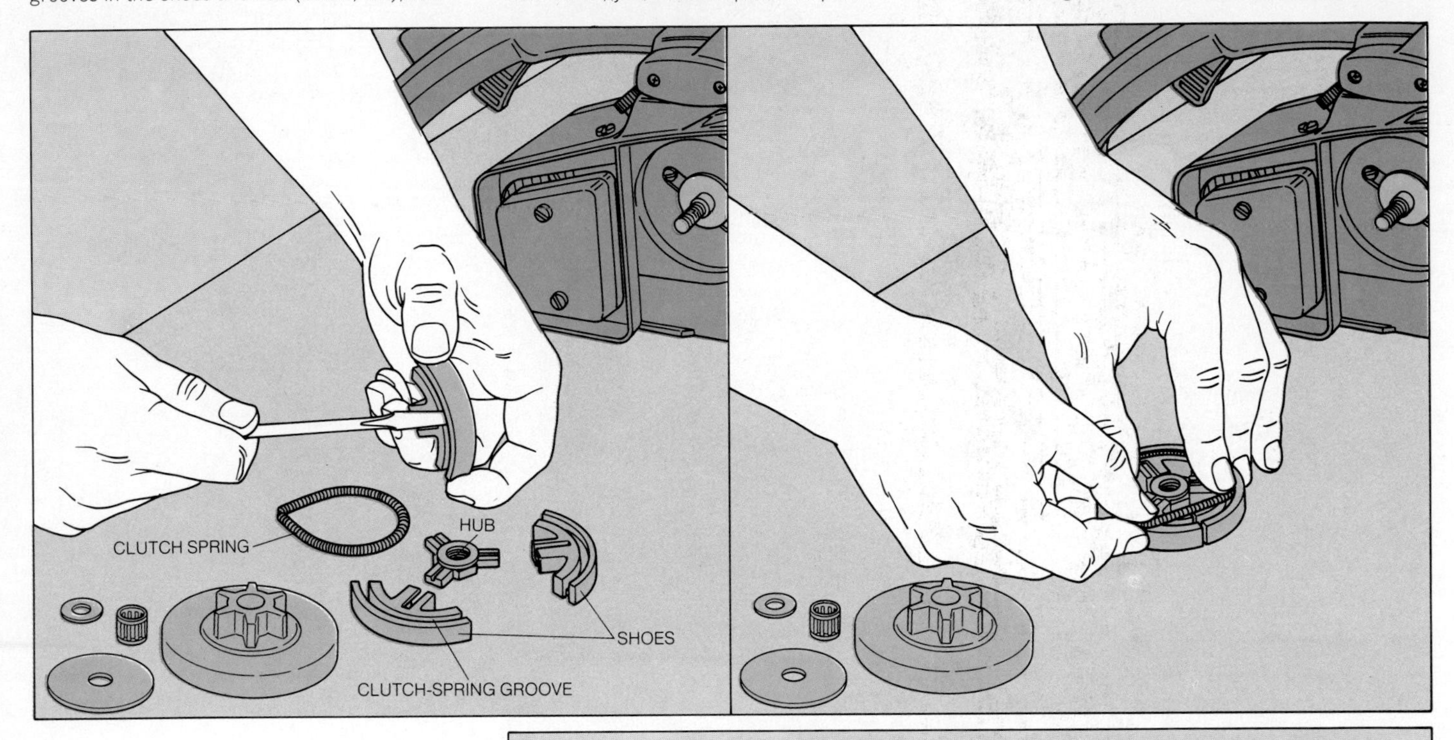

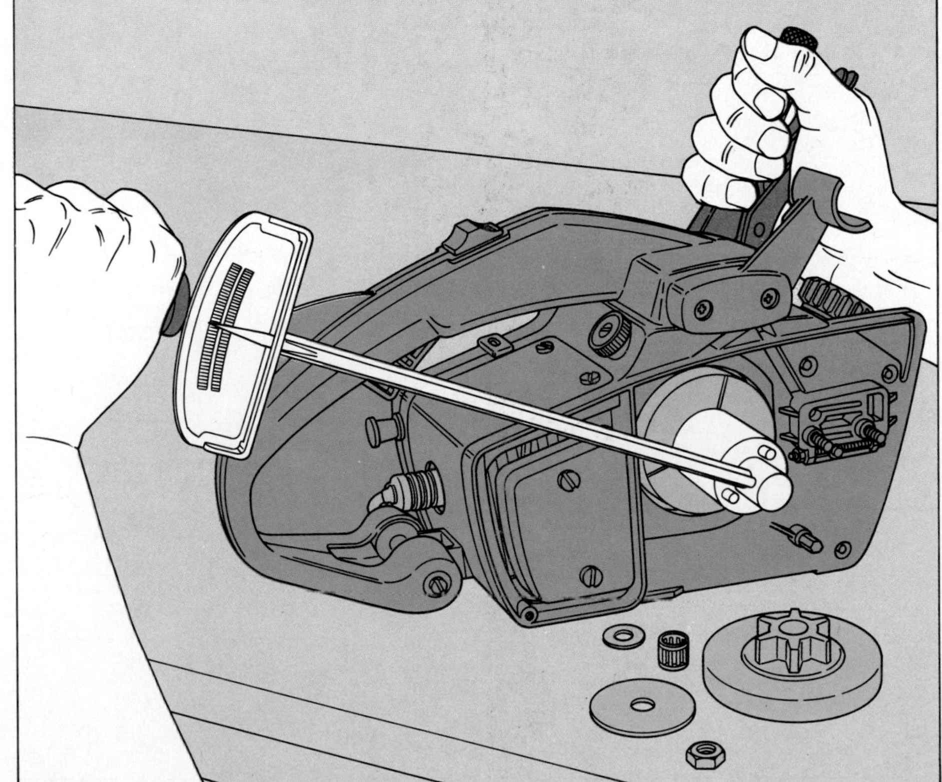

3 Putting the clutch assembly back on. Thread the clutch onto the crankshaft with the spring facing in and, using the clutch wrench and a torque wrench, twist the clutch counterclockwise, tightening it to the torque specified in the owner's manual. Replace the rest of the parts in order. Grease the needle bearing lightly with a lithium-based grease before replacing it; do not get grease on the clutch shoes. Once the clutch assembly and drive sprocket are back on, replace the guide bar and chain, the guide-bar cover, the fan housing and the front hand grip.

Adjusting the Oil Pump

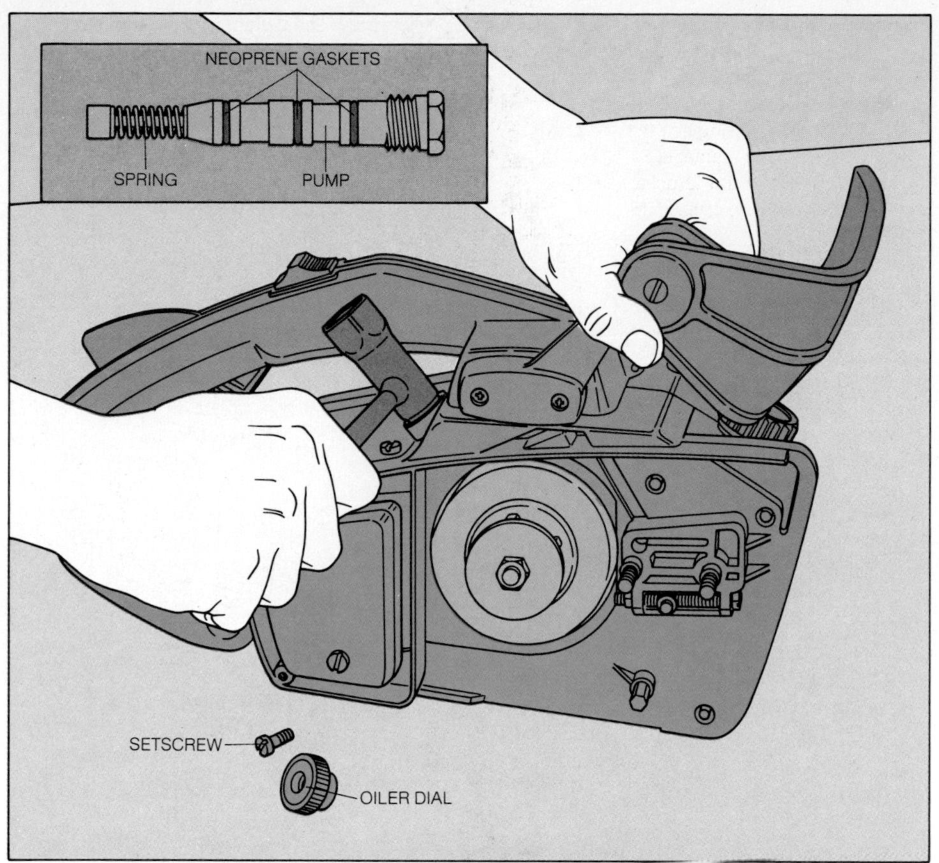

1 Removing and inspecting the oil pump. After removing the guide-bar cover, the guide bar and the chain, loosen the setscrew at the center of the oiler dial; lift out the screw, the dial and the washers below. Use a wrench to twist the oil pump loose, then lift out the pump. Dip all of the parts *(inset)* in kerosene to clean them, then inspect the neoprene gaskets on the pump and replace any that are cracked or swollen. If the spring at the base of the pump is broken or worn, replace it.

Wipe a light coat of chain oil over the gaskets, then push the pump back into place. Tighten it with a wrench until it is just snug, then loosen it three quarters of a turn.

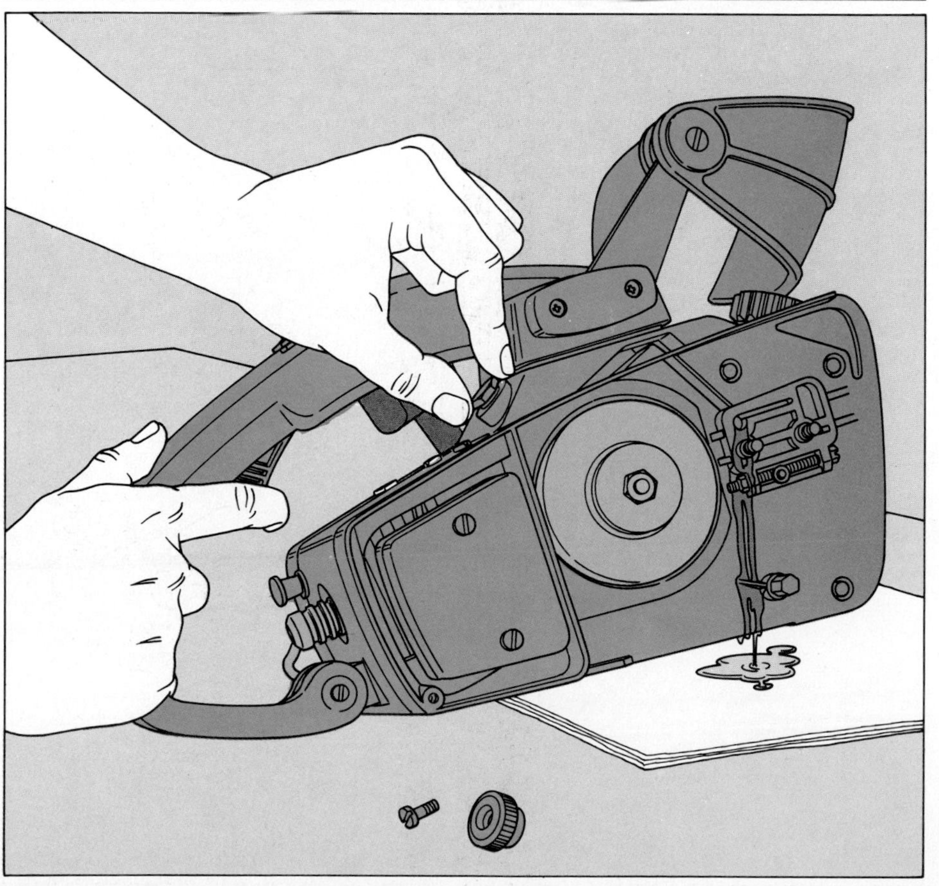

2 Adjusting the oil flow. Free the oil port of accumulated oil and sawdust and, holding the saw over several layers of newspaper or a dropcloth, start the engine and depress the throttle trigger. Be sure not to run the saw at more than half its normal speed; without the guide bar and chain attached, high-speed operation can damage the engine. While the saw is running, oil will flow freely from the oil port onto the newspaper; with the engine still running, slowly twist the head of the oil pump counterclockwise with your fingers or a wrench until the oil flow is a small but steady trickle. Then turn off the saw.

Replace the washers, the dial and the setscrew; position the dial and its catch so that the dial will turn clockwise only. Then reattach the guide bar and chain, and adjust the chain tension *(page 102, Step 4)*. Hold the nose of the guide bar 5 inches above the newspaper and run the saw at full speed; while the engine is running, turn the oiler dial until a slow, steady stream of oil drips off the guide bar.

The Rotary Mower: A Work Horse for the Lawn

The rotary lawn mower exploits the power of the small gasoline engine in the simplest way possible. The long, flat cutting blade is connected directly to the end of the crankshaft; when the engine runs and the shaft turns, the blade spins.

The design means that the standard rotary mower is easy to maintain. An annual tune-up *(pages 38-53)* and periodic care during the mowing season will ensure years of trouble-free service.

The most common cause of premature aging in a lawn mower is rust or corrosion, and cleaning is a crucial part of routine maintenance. After each use, wipe moisture-holding dirt off the mower body and scrape accumulated grass and soil from the underside of the housing. Also check the mower for rust. Buff any rust spots away immediately with steel wool, and cover the exposed metal with touch-up paint—available at lawn-mower-repair shops or hardware stores. Finally, check the top of the mower housing for streaks of grease from the engine; remove them with a commercial degreasing solvent.

After every 20 hours or so of operation, a lawn mower requires a more thorough going over—the bolts and control cable should be checked and lubricated and the blade serviced. Bolts in the mower handle and housing are loosened by the vibrations of normal operation, and need only be cleaned off, tightened and then coated with oil to protect them from corrosion.

After long use, the control cable, which governs the flow of fuel into the carburetor and thus the speed of the engine, may become stiff and unresponsive. The cable runs from a lever on the mower handle down to a throttle control lever at the carburetor; it consists of a stiff inner cable enclosed in a protective outer casing. To keep it functioning smoothly, simply trickle oil down through the casing *(opposite)*.

Servicing the blade requires a little more work. A mower blade is attached to the crankshaft with a single bolt, which is likely to be frozen and extremely difficult to remove. The best way to loosen the bolt is with a hammer and an impact driver *(page 16)*—although an open-end wrench or a socket wrench applied with vigorous force can sometimes do the job.

Nicks or gouges in the blade can make one side lighter than the other—setting up vibrations that not only will prematurely loosen bolts but can cause serious engine damage. The simplest way to check blade balance is to weigh the blade on an inexpensive spike-and-cone blade balancer, available at a hardware store. Once you have determined which is the heavier side, you can compensate when sharpening the blade—either on a grinding wheel or with a metal mill file.

Eventually, a blade can become so narrowed through repeated sharpenings that it is no longer effective and should be discarded. As a rule of thumb, a blade should be replaced if it has lost about a quarter of its original width at the ends.

When tipping a mower over to clean it or to work on its blade, take care not to flood the carburetor. If the mower is equipped with a fuel shutoff valve, close it. Otherwise, make sure that the carburetor stays above the fuel tank. If the engine will not start after it is righted, the carburetor has flooded nonetheless and must be drained *(pages 54-55)*.

To prevent gas from dribbling out through the vent on the fuel cap, cover the cap with a piece of plastic secured with a string or a rubber band.

Most important of all, before tipping the mower or doing any work on the blade, engine or housing, make sure that the engine cannot start suddenly. Because the blade spins directly on the crankshaft, inadvertently rotating it can start the engine as effectively as pulling the starter cord. The only way to immobilize the engine safely is to remove the spark-plug wire from its plug. The wire must then be secured so that it cannot swing back anywhere near its connected position, because an electric spark can jump small gaps.

Some mowers are equipped with a clip that holds the wire away from the spark plug. On others, attach the wire firmly to the mower body with a piece of electrical tape before doing any work.

An additional hazard is that rotary mowers customarily are equipped with clutchless blades: As long as the engine is operating, the blade will turn. Often, momentum will keep it twirling after the engine has stopped. Because of the dangers inherent in such a design, federal regulations that went into effect in 1982 require mowers to be manufactured with a clutch to disengage the blade from the engine and brake it within three seconds after the control lever on the handle has been released.

Working On the Exterior

Tightening and lubricating the bolts. Wipe any accumulated grime off the bolts in the mower handle. Then tighten the bolts with a wrench, and coat them with a film of lightweight machine oil to protect them from corrosion. Tighten and oil the pivot points of the height-adjustment levers and the axle pivots on the wheels.

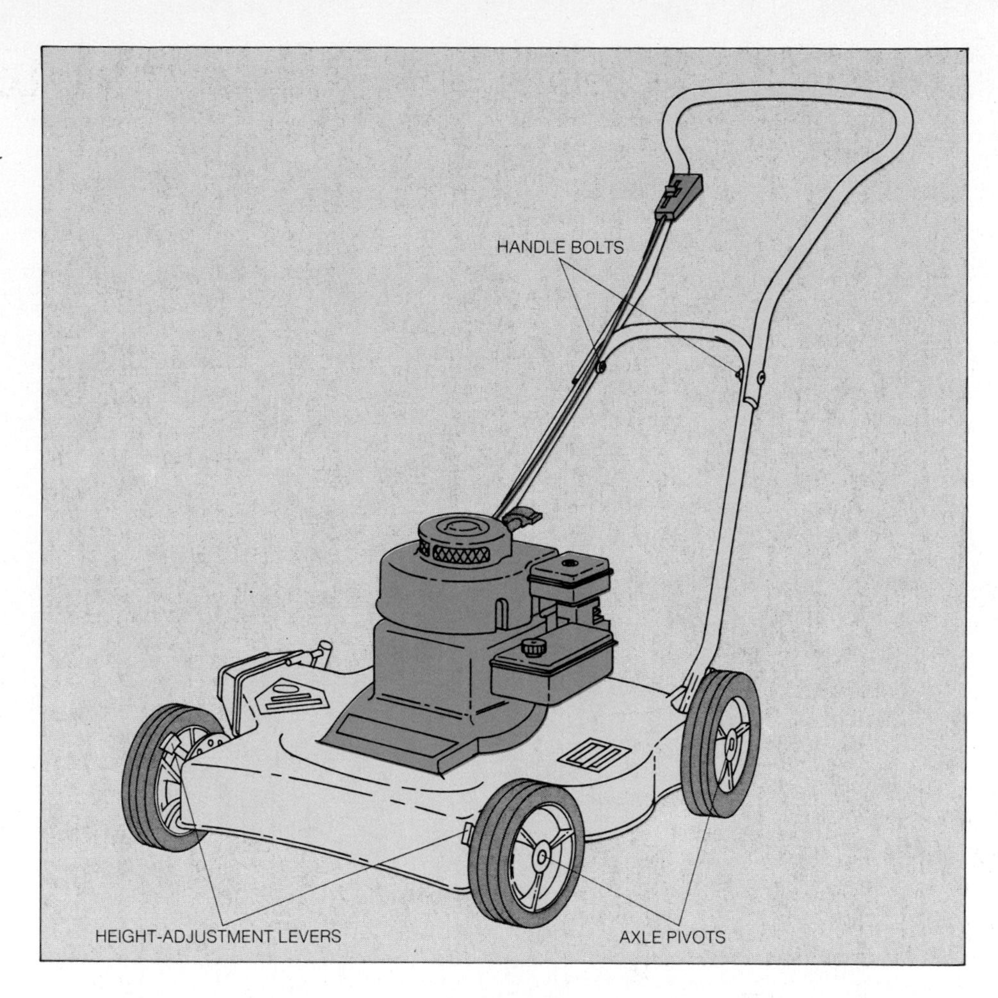

Caring For the Cables

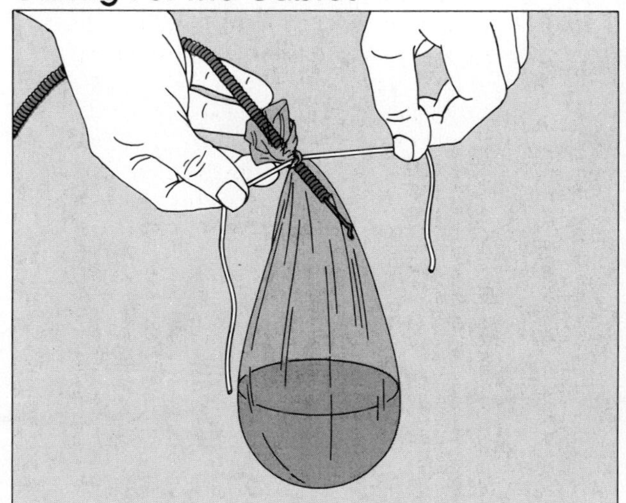

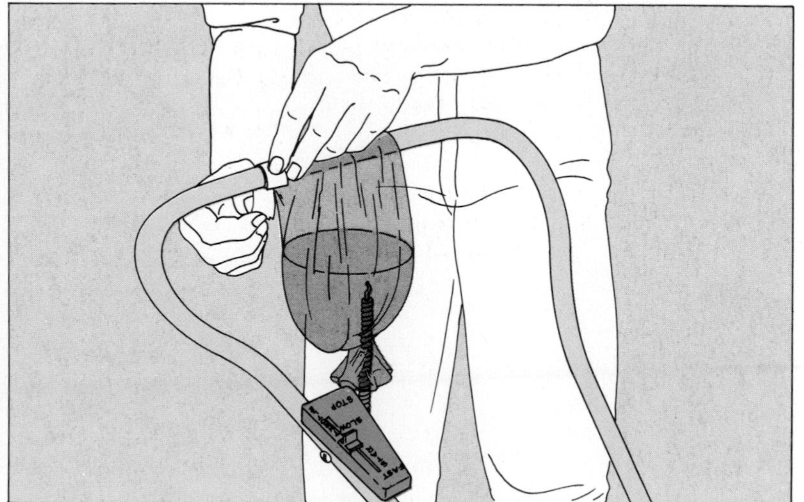

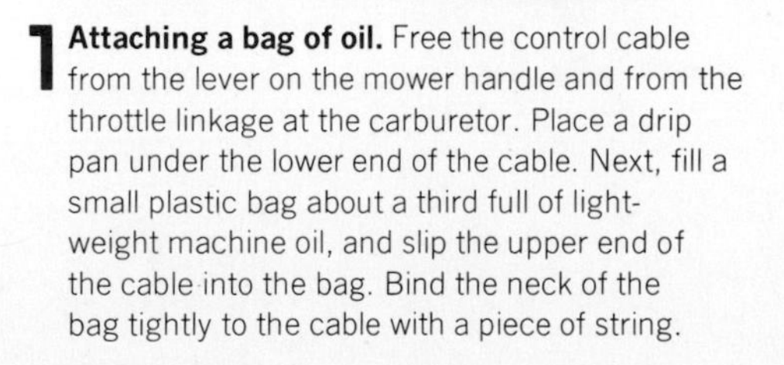

1 Attaching a bag of oil. Free the control cable from the lever on the mower handle and from the throttle linkage at the carburetor. Place a drip pan under the lower end of the cable. Next, fill a small plastic bag about a third full of light-weight machine oil, and slip the upper end of the cable into the bag. Bind the neck of the bag tightly to the cable with a piece of string.

2 Infusing the oil. Use a strip of electrical tape to hang the oil-filled bag from the mower handle. When oil starts to flow from the lower end of the cable into the drip pan—after about five minutes or so—remove the bag.

3 Reattaching the cable. After hooking the upper end of the control cable to the control switch on the mower handle, slide the switch to STOP. Slip the lower end of the cable through the cable clamp near the carburetor. Then hook the inner cable to the throttle control lever and pull the cable taut back through the clamp so that the lever is in the stop position. Tighten the screw on the clamp to hold the cable in place.

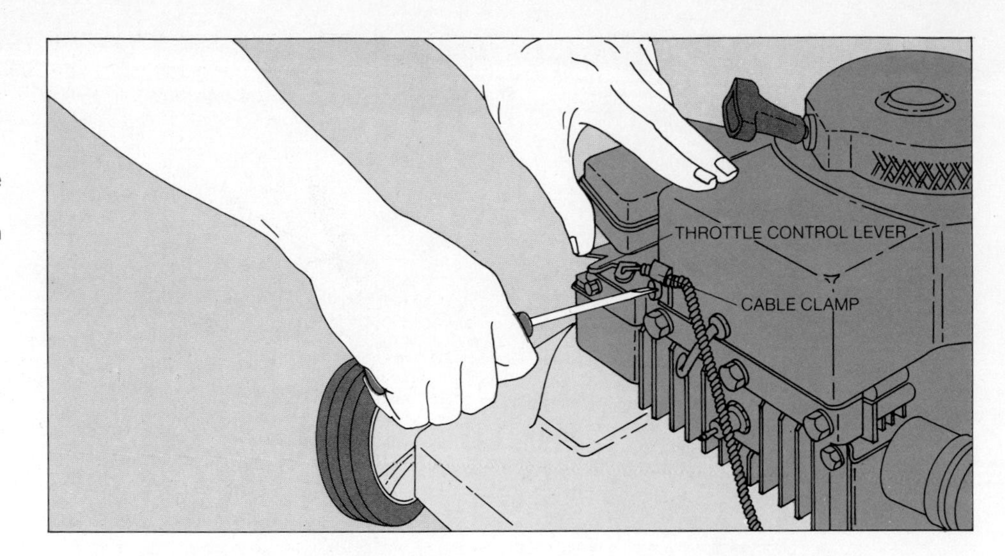

Servicing a Blade Assembly

1 Removing the blade. After disconnecting and securing the spark-plug wire, tip the mower on its side and squirt penetrating oil around the edges of the blade bolthead. Wait about five minutes for the oil to seep into the bolt threads. Then, wearing heavy canvas gloves, loosen the blade bolt with an impact driver set to turn in a counterclockwise direction *(page 16)*. Remove the bolt with a wrench. If you do not have an impact driver, immobilize the blade by wedging a block of wood between it and the housing and unscrew the bolt with an open-end wrench or a socket wrench.

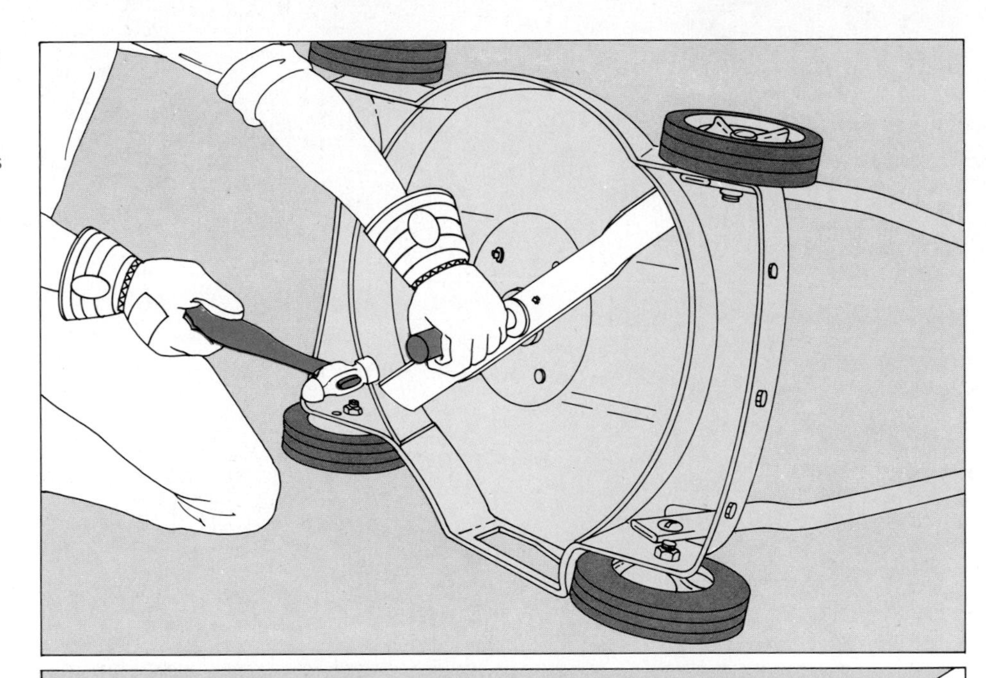

2 Checking blade balance. Scrub the blade thoroughly clean with a wire brush. Place the cone of a blade balancer *(inset)* on its supporting spike. Fit the bolthole in the blade over the tip of the cone. If the blade is out of balance, it will tilt seesaw-like toward the heavier side.

If you do not have a balancer, slip the blade onto the shaft of a small screwdriver. Hold the screwdriver horizontal; if the blade is out of balance, it will turn like a propeller, the heavier end coming to rest pointing downward.

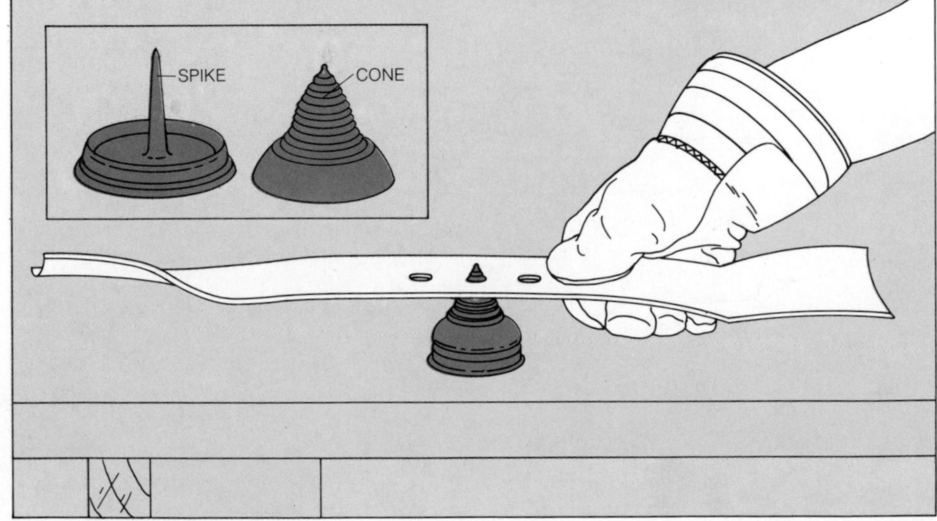

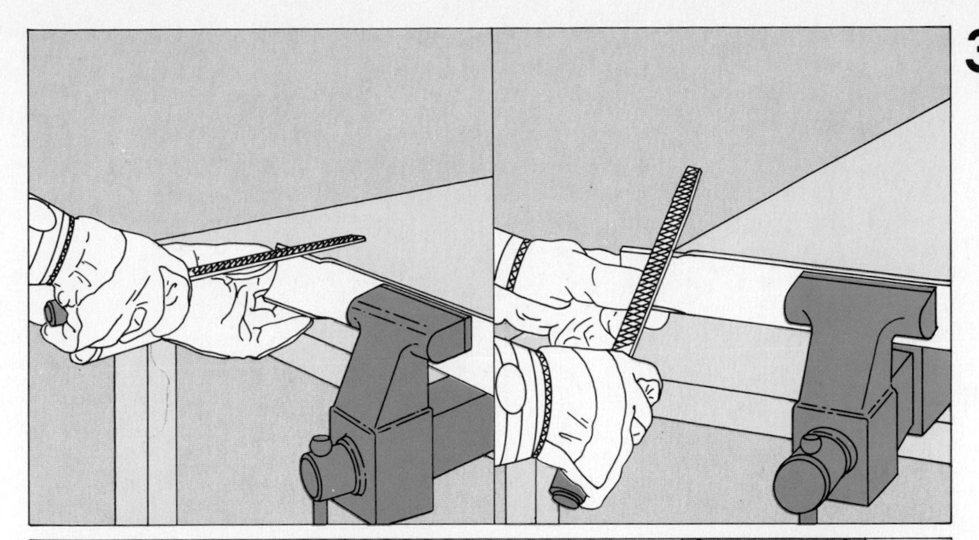

3 Sharpening the blade. Clamp the blade in a vise, then use a mill file drawn straight across the edge to dull it and remove any nicks *(far left)*. When the blade is smooth, resharpen it by filing along the original angle of the edge *(left)*; move the file away from you, in one direction only. Take a few extra strokes of the file on the heavier end to lighten it. Do not try to make the blade razor-sharp by filing the flat side of the edge; a too-thin edge is vulnerable to nicks and chips.

After sharpening both ends of the blade, recheck the balance. Do not continue to file the sharpened edge if the blade is still unbalanced. Instead, lighten the heavier end by taking a few strokes of the file across the blunt side opposite the sharpened edge.

4 Removing the blade adapter. Use a plastic-tipped mallet or the handle of a hammer to tap the blade adapter off the end of the crankshaft. (The adapter is connected to the shaft with a small metal key, which may drop loose as you remove it.) Examine the small lugs that fit into the holes on either side of the bolthole. If the lugs are damaged, they will not grasp the blade firmly and the adapter should be replaced.

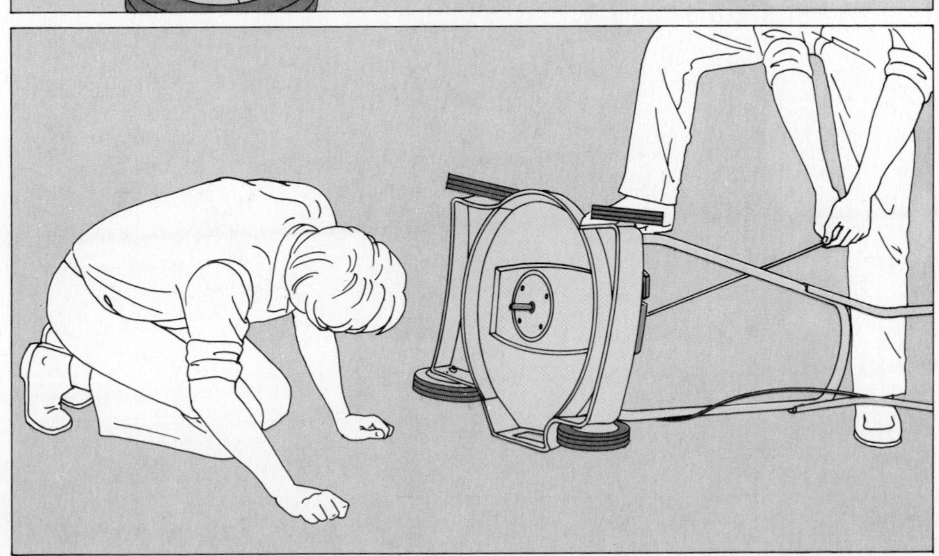

5 Checking the shaft. To see whether the shaft is straight, sight down it while a helper pulls the starter cord. An absolutely straight shaft, when viewed head on, will not appear to move at all as it spins. If it vibrates or wobbles, it is bent and must be replaced *(pages 72-79)* before the blade is reinstalled.

Before remounting the blade, use a socket wrench to tighten the bolts holding the engine to the mower body. Then fit the adapter and blade onto the shaft, screw on the blade bolt and tighten it with an impact driver or a socket wrench. Tighten it as firmly as possible; there is virtually no danger of stripping threads on the large blade bolt. If you have a torque wrench, tighten the bolt to 50 foot-pounds.

Mowers That Propel Themselves

Self-propelled rotary mowers have simple belt- or chain-driven transmissions that relay power from the engine to the wheels. On a belt-driven mower, the belt loops around two pulleys. One, the drive pulley, is connected directly to the crankshaft. The other is mounted on the gearbox, which contains a clutch mechanism that transfers the power of the running belt to the wheel axle.

A chain-driven mower operates in similar fashion, but the chain is looped around two toothed sprockets, and the drive sprocket is connected to a separate power-take-off shaft.

A self-propelled mower requires the same maintenance as a push-propelled one, with the addition of a few extra chores. When you lubricate the bolts and cables, give a light coat of bearing grease to the gears behind the hubcaps on the powered wheels. At the same time, lubricate the pulley or sprocket shafts with lightweight machine oil. Do not let oil reach the belt: A greasy belt will slip.

On a belt-driven mower, check the condition of the belt to see if it has cracked or hardened. Adjust the belt tension *(below)* if necessary.

With a chain-driven mower, check the chain after every three or four uses. If it is dry, spray it with an aerosol chain lubricant. If the chain is loose or if any links are bent or broken, replace the chain *(opposite, top)*. Never try to repair or shorten it by removing links. If any teeth on either of the sprockets are damaged, replace that sprocket *(opposite, bottom)*.

The self-contained sliding-fork gearbox typical of most self-propelled mowers should give years of trouble-free service without special care. At times the wheels will suddenly fail to turn, though, indicating a possible problem in the gearbox.

Before dismantling the gearbox to find the problem *(pages 112-113)*, disconnect the spark-plug wire, remove the chain or belt cover and pull on the starter cord to see if the gearbox sprocket or pulley turns. If it does not, the problem is a slipping belt or—on a chain-driven mower—something amiss with the take-off shaft. Also check the clutch control cable to make sure it is pulling the shift arm at the gearbox. Finally, remove the hubcap on one of the powered wheels, engage the clutch control cable and pull the starter cord once more. If the pinion gear at the end of the axle does not rotate, the problem is in the gearbox.

A Belt-driven Mower

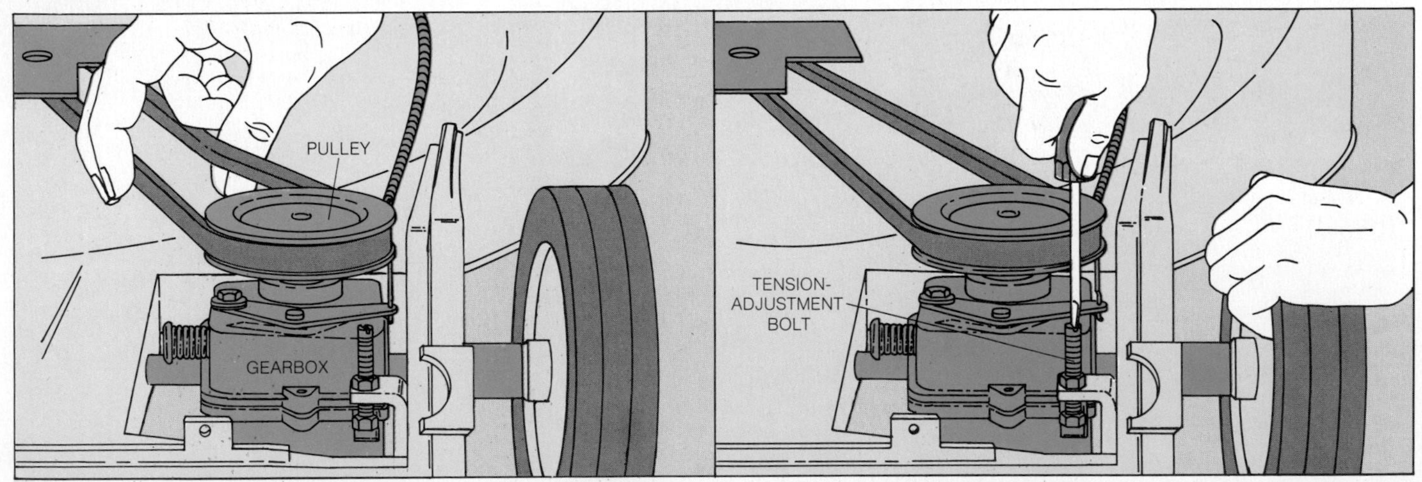

Examining the belt. Turn the mower off, disengage the spark-plug wire and remove the belt cover. Depress the belt midway between the pulleys with your fingers *(above, left)*. If the belt gives more than half an inch, it is too loose. Tighten the bolt that holds the gearbox to the mower housing *(above, right)*. This will pivot the gearbox slightly, moving the pulley attached to it and increasing tension on the belt. Some mowers are equipped with a third, unpowered pulley called an idler pulley, which presses against the belt to maintain the correct tension. Move the idler pulley to tighten the belt.

If the belt's inside surface is smooth and shiny, the belt has glazed—become too slick for the pulleys to grasp. Spray it with a deglazing compound from a lawn-mower-repair shop or auto-supply store. If the belt is cracked or hard, replace it. To remove it, loosen the tension and slide the belt off its pulleys. Tip the mower onto its side, reach up along the crankshaft and pull the belt down through the housing. On some mowers, you may first have to remove the blade.

A Chain-driven Mower

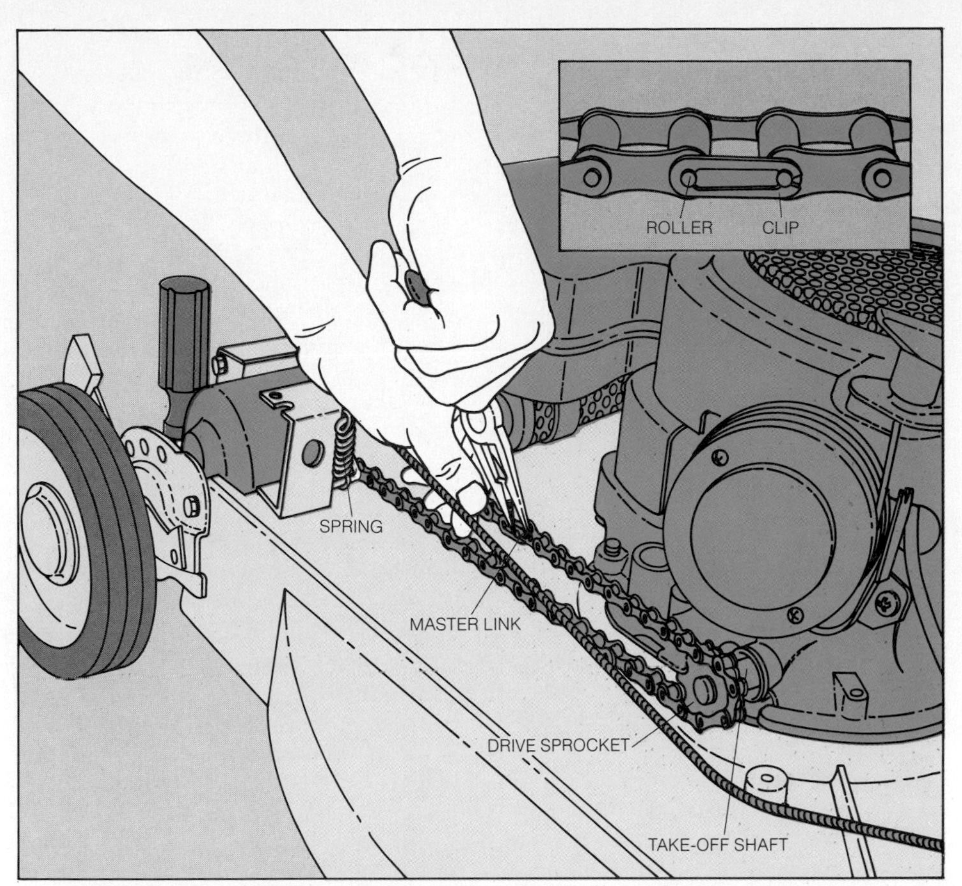

Replacing the chain. Chain tension usually is maintained by a spring that pulls the gearbox away from the auxilary power-take-off shaft, which is connected to the drive sprocket. To remove the chain, disengage the spark-plug wire, then wedge a screwdriver behind the gearbox to tilt the box toward the take-off shaft. If necessary, rotate the chain by pulling on the starter cord until the master link *(inset)* is accessible. Disconnect this link by prying its clip off the rollers with a pair of pliers.

Slip the new chain over the sprockets and connect its ends with the master link, making sure that the open end of the clip faces away from the direction in which the chain turns. Otherwise, the clip may snag on a sprocket and come off.

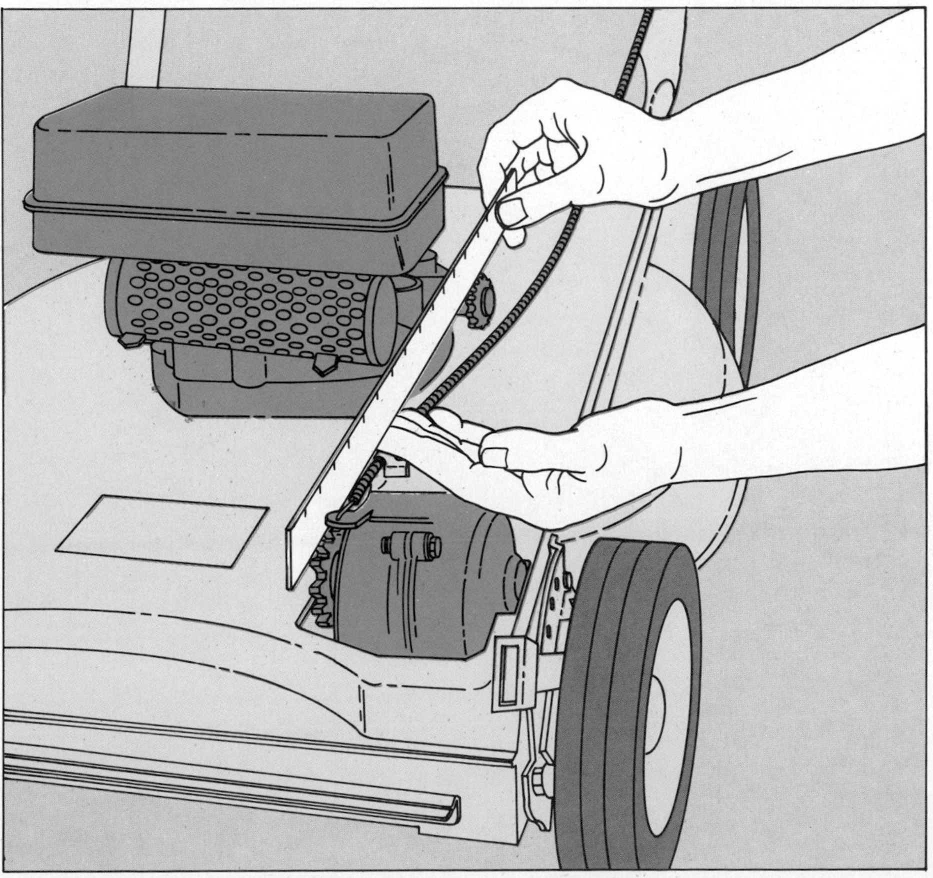

Replacing a damaged sprocket. Use a hex wrench to remove the setscrew that holds the sprocket to its shaft. Slide the sprocket off the shaft and slip a new one into position. Check to make sure it is aligned with the other sprocket by holding a steel ruler over the tops of both. When they are on the same plane, tighten the setscrew on the new sprocket.

Troubleshooting the Transmission

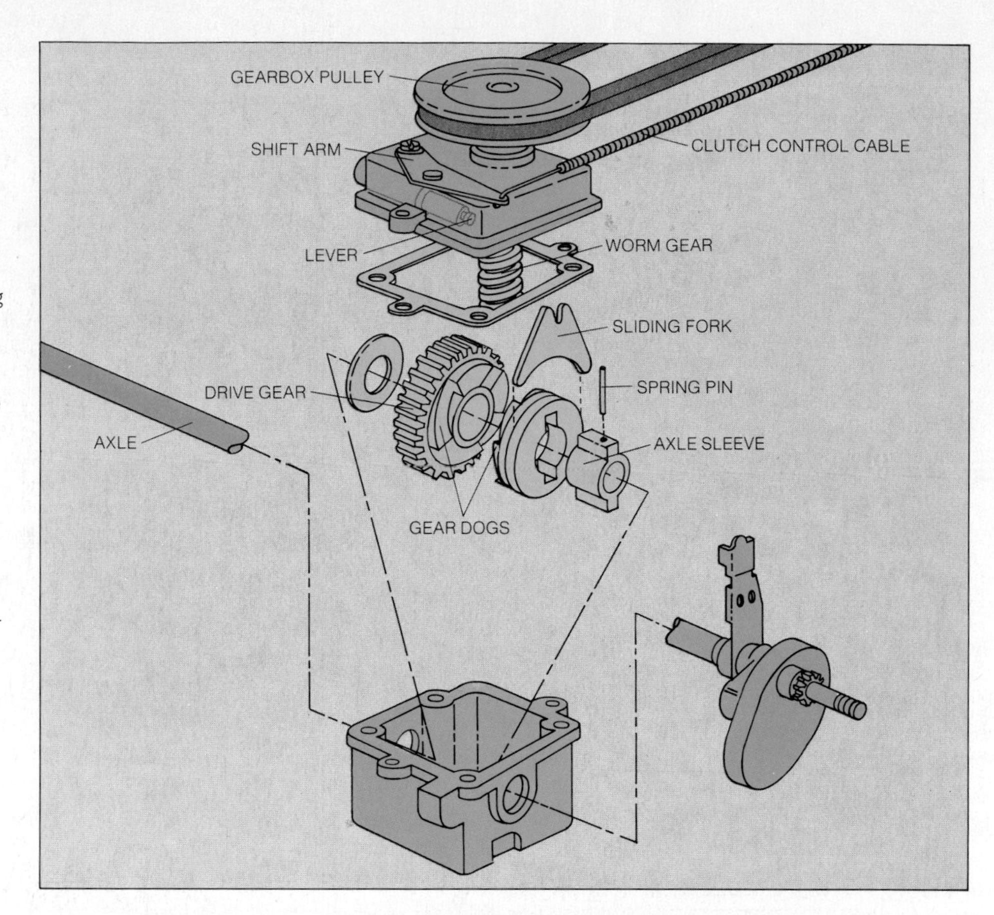

Anatomy of a sliding-fork gearbox. The gearbox pulley spins a worm gear, which in turn rotates the drive gear on the axle. A ridged axle sleeve is attached to the axle with a spring pin. The sleeve is encircled with a clutch jaw that slides back and forth along the ridges. The sliding fork, from which the assembly gets its name, rests in a groove on the rim of the clutch jaw.

When the clutch control cable is pulled to engage the wheels, the shift arm pulls a lever that tugs the sliding fork toward the drive gear, pulling the clutch jaw with it. The projecting gear dogs on the clutch jaw then mesh with those on the drive gear, and the axle turns.

Although any of the components may become damaged, the commonest problem with a sliding-fork gearbox is broken or worn gear dogs. If any are damaged, both the drive gear and the clutch jaw must be replaced.

Overhauling the Gearbox

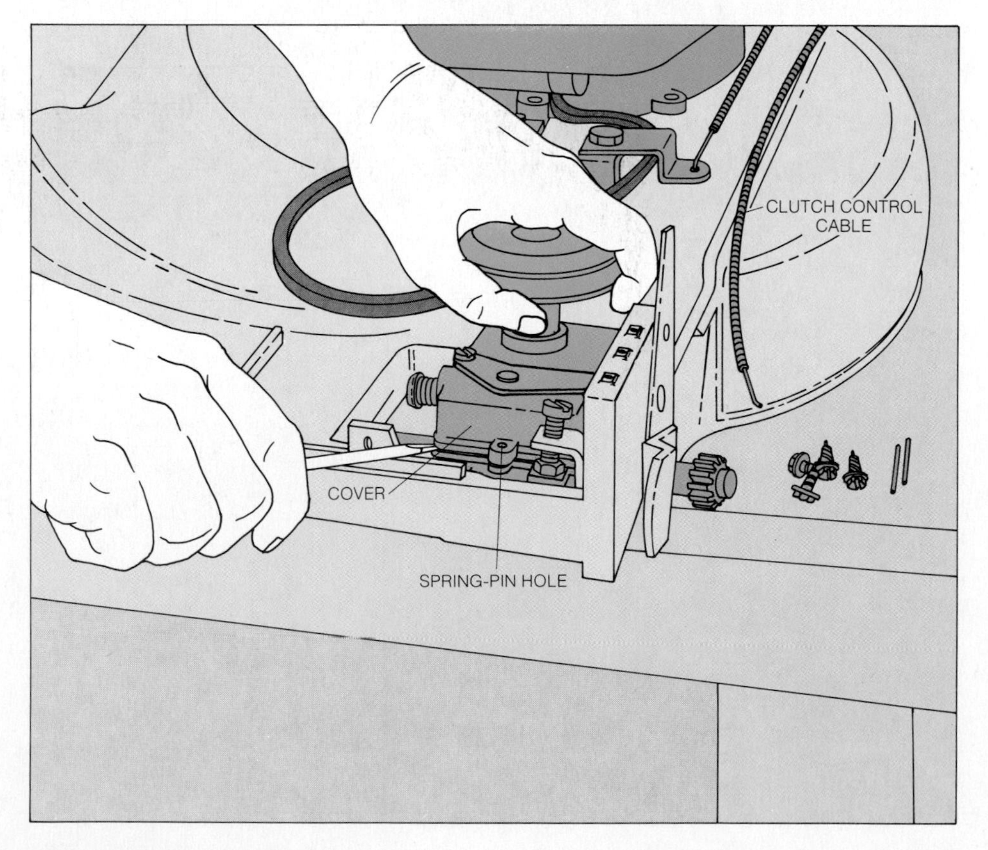

1 Removing the cover. Slip the belt or chain off the pulley or sprocket on the gearbox, and unhook the clutch cable from the shift arm. Remove the wheel nearest the gearbox, then tap out the spring pins at the front and back of the box. Use a socket wrench to remove the screws that clasp the two halves of the box together. Place a drip pan under the box, and pry off the cover with an old screwdriver or a pry bar.

2 Removing the axle spring pin. Rotate the axle so that the top of the spring pin in the axle sleeve faces upward. Use a hammer and a pin punch to tap the pin downward. Then rotate the axle so that you can reach the protruding end, and pull the pin out with pliers.

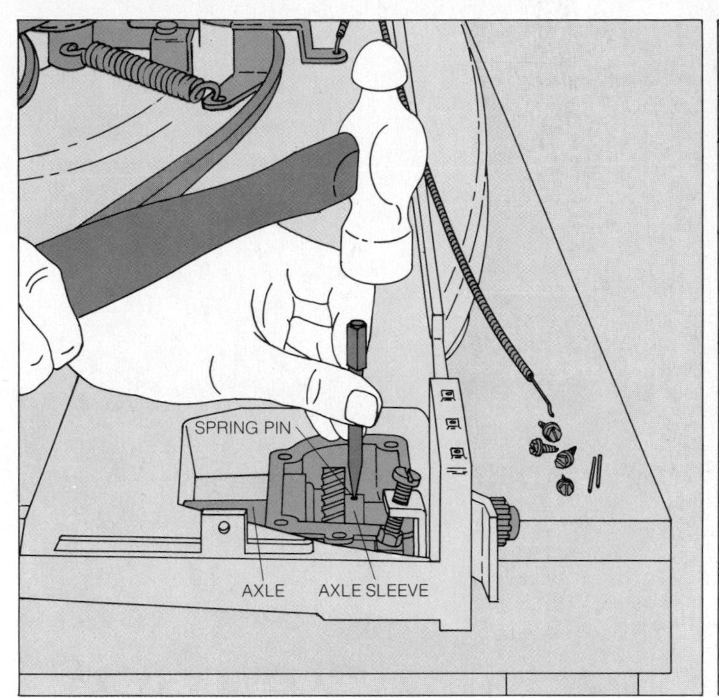

3 Removing the pinion gear. Use a ball-peen hammer and a pin punch to tap out the spring pin that fastens the pinion gear onto the end of the axle. Pull the gear off the axle.

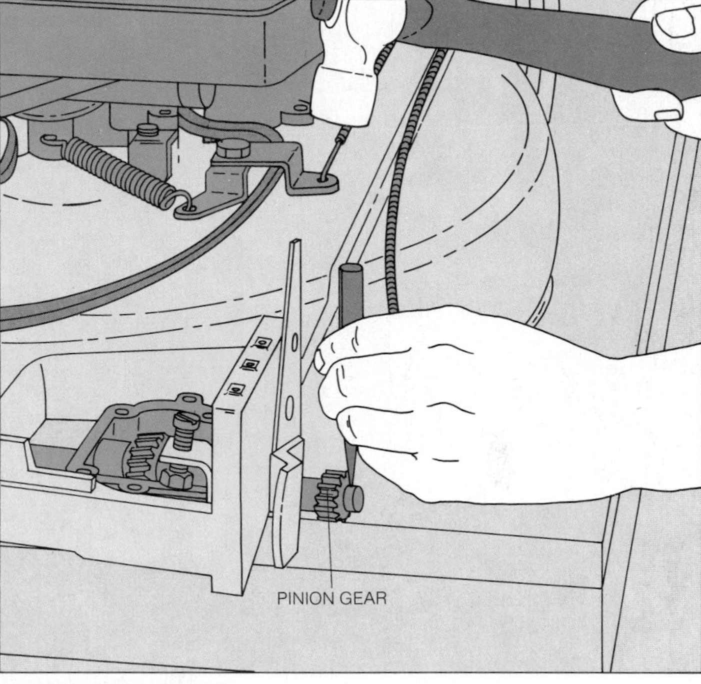

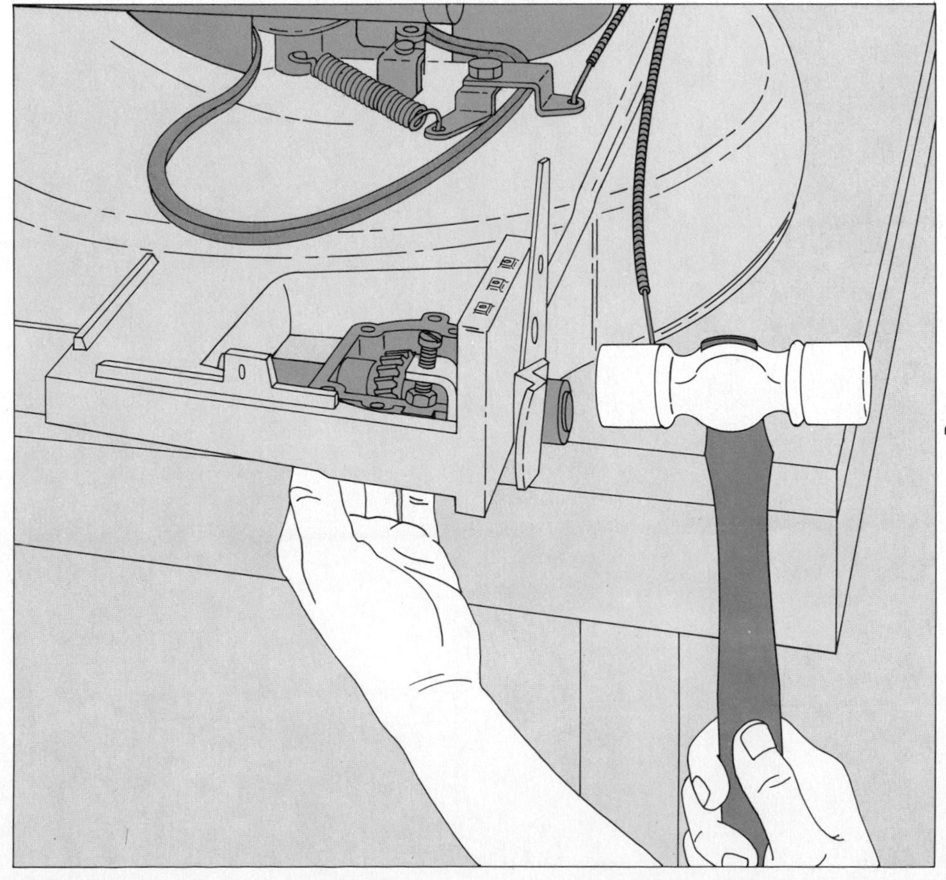

4 Examining the gearbox. Using a plastic-tipped mallet, tap the end of the axle through the bearing in the mower housing, and slide the gear case off the end of the axle. Carefully examine the drive gear, sliding fork, clutch jaw and sleeve; replace any damaged parts. Reassemble the gearbox by fitting all of the parts back into the case; slide the axle through them. Tap the axle into position through its bearing. Then align the spring-pin holes in the sleeve and the axle, and tap in the pin.

Fill the box two-thirds full with fresh SAE 90-grade transmission oil. Then attach the cover to the box with its spring pins and screws, tightening the screws to 60 to 70 inch-pounds with a torque wrench. Attach the pinion gear to the axle, and install the wheel and hubcap.

Laborsaving Riding Mowers for Big Yards

Riding mowers and lawn tractors make quick work of the verdant expanses that once represented pocket money to industrious schoolboys. Today, with less effort than it takes to drive a car, the owner of one of these jaunty machines can trim a lawn as large as an acre in as little as 30 minutes. Best of all, the machines require only minimal attention to repay their owners with years of hard-working service.

Riding mowers and lawn tractors differ primarily in the size and placement of their engines. Riding mowers are driven by 4- to 6-horsepower, rear-mounted engines, while lawn tractors sport larger 8- to 12-horsepower engines, mounted in front. Both, however, have identical drive trains and blade-control mechanisms. Although various makes and models may differ slightly from one another—some boast five forward speeds plus a reverse gear, while others have only two forward speeds and no reverse—the system of belts, pulleys, chains and sprockets found in any mower or tractor will be similar. For any extensive maintenance or repair work, however, you may find it helpful to supplement the instructions on these pages with the owner's manual or service manual for your particular model.

Much of the required maintenance on riding mowers and lawn tractors is aimed simply at keeping them clean. The air filter should be cleaned after every 10 hours of operation—even more frequently in dusty conditions—and the oil should be checked and changed at regular intervals. Additionally, the engine should be maintained and tuned in the same way as any small engine *(Chapters 2 and 3)*.

Maintenance of the nonengine parts of the machine should begin with cleaning leaves, twigs and other debris from beneath the machine after every use to ensure an unobstructed path for the cutting blade. The belts that drive the blade must be kept clean and dry as a safeguard against premature wear, the blade must be kept sharp and balanced *(pages 108-109)*, and the chains that transmit engine power from the drive mechanism to the wheels should be inspected often for lubrication and correct tension. Also, regular lubrication of all friction points on the frame and linkages will keep the moving parts running smoothly.

Although a periodic inspection of the mower's belts and chains should reveal excess wear in time to prevent a breakdown, often damage or wear is concealed and the first sign of impending trouble is a rough or erratic ride. Depending on the symptom, you may be able to pinpoint the problem without dismantling the mower. In many cases, however, you must disassemble parts of the drive train—in the sequence of the steps shown on the following pages—to locate the source of the trouble and replace or repair the worn parts.

If the mower loses traction, a chain may be loose or broken, requiring tightening or replacement. Loss of traction may also mean that the drive disk, the aluminum plate fixed to the crankshaft *(opposite)*, has worn smooth, causing it to slip. You can easily restore the necessary roughness to the finish by rubbing the disk with sandpaper or emery cloth.

If the drive mechanism slips or if you cannot get the machine to stay in gear, it is probable that one or more of the linkage rods are out of adjustment. In most cases, these rods are anchored at one or both ends by a slotted plate or other device whose position can be shifted *(page 119)*. A loss of neutral gear, for example, means that the rod that controls the vertical travel of the drive wheel—the rubber-tired roller perpendicular to the drive disk—needs adjustment. A loss of reverse gear, or slippage of the highest gear, requires an adjustment of the rod that controls the horizontal travel of the drive disk.

A loss of control over the blade may require a number of adjustments. If the blade does not turn, either of the two belts connecting the blade to the engine may be damaged or the blade nut may be loose. If the blade continues spinning for more than six seconds after you have set the blade brake, the brake or rear blade belt needs adjustment. Telltale swirl marks left on the lawn after cutting indicate that the blade pan, the housing that contains the blade, is misaligned. To ensure that the grass is cut only as the blade passes the front of the blade pan, adjust the tilt of the pan so that the front edge is ¼ inch lower than the rear edge.

Unless your mower or tractor has an unusually high ground clearance, you will sometimes have to jack it up or tilt it on end to service parts underneath the machine. Riding mowers are designed to be lifted and rested on a tilt bar, built into the rear end of the machine. Most lawn tractors must be raised on one side with a jack to expose their underside. Check your owner's manual for the safest places to put the jack on your model.

Before lifting or jacking a mower or tractor, drain both the fuel tank and the oil reservoir *(page 27)*; if your mower is an electric-start model, remove the battery as well. When standing the mower on end, as with a tilt bar, always tie it securely to a wall or a door jamb to keep it from tipping or falling. Place the throttle control in the stop position and disconnect the spark-plug wire before starting any inspection or repair.

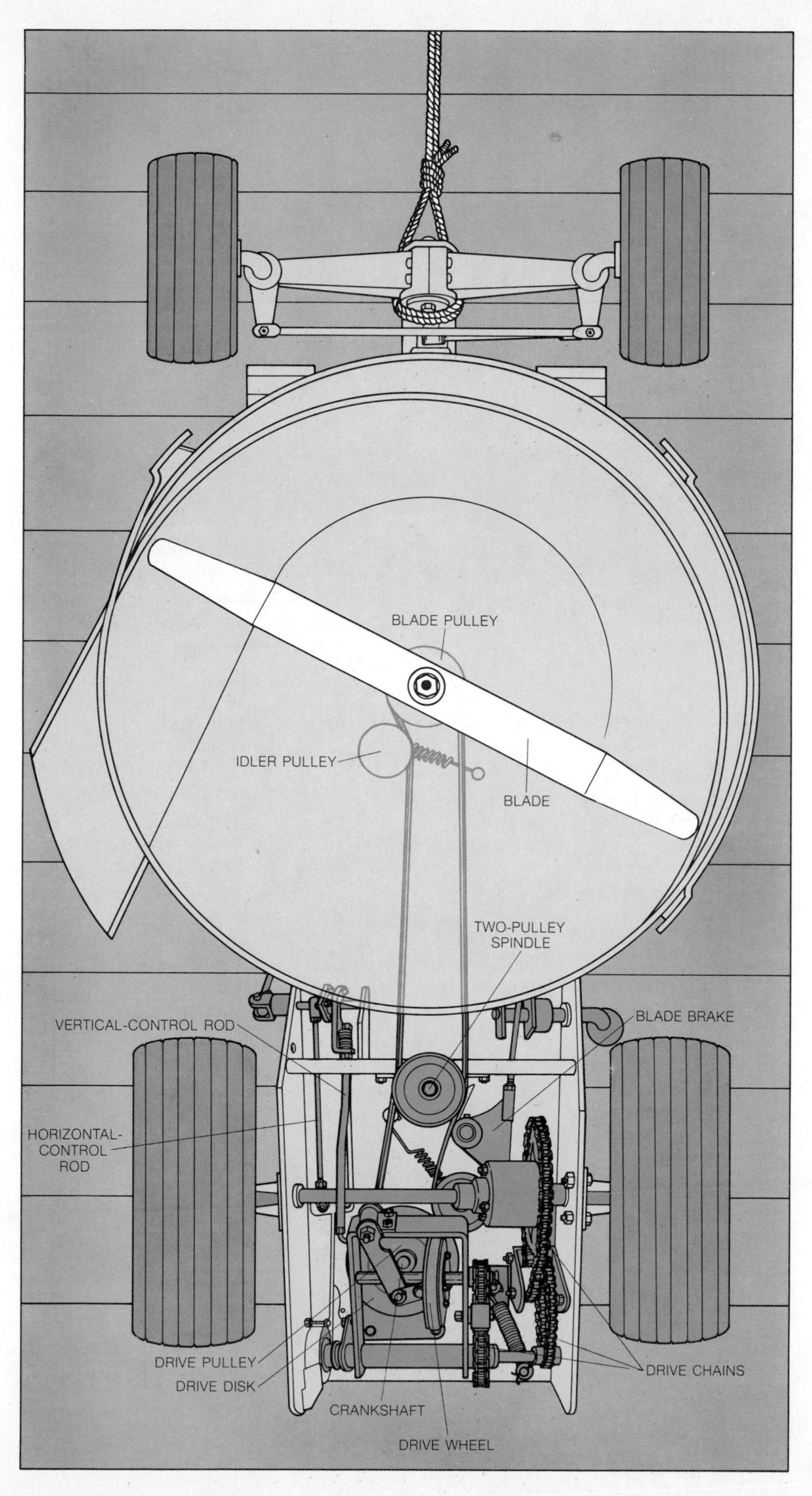

The Plates and Pulleys That Turn the Wheels

The drive and blade systems. The rotary motion of the spinning crankshaft is transferred to the rear wheels by an aluminum plate, called the drive disk, mounted on the crankshaft. As it rotates, it turns the drive wheel, a rubber-rimmed wheel set perpendicular to the disk. The drive wheel then transmits its motion through a series of sprockets and chains, the last of which turns the rear wheels of the mower. The speed and direction of the drive-wheel rotation, and therefore of the mower, are determined by where the drive wheel contacts the drive disk, which can be altered by moving the mower's shift lever *(page 116)*.

A system of rubber belts and pulleys spins the cutting blade. A drive pulley, positioned directly above the aluminum drive disk, is keyed to the crankshaft. A short belt runs from this pulley to the top pulley of a two-pulley spindle, from which a long belt connects to a pulley on the blade shaft. A belt guide and a spring-loaded idler pulley keep the short belt tight. The long belt is tightened and loosened by an idler pulley whose pressure engages or disengages the blade. This idler pulley, in turn, is operated by a lever on the steering column. A blade brake, which is activated automatically when the blade is disengaged, stops the blade in seconds by applying pressure to the short belt.

Shifting the drive-wheel position. The drive system—seen here from the rear with the mower on the ground—produces varying mower speeds from a single engine speed by bringing the drive wheel into contact with the drive disk at varying distances from the center of the disk. The drive wheel turns fastest when it is touching near the rim of the spinning drive disk *(top)*, slower when it is moved toward the center of the drive disk. When the mower is shifted into reverse, the drive wheel travels across the center of the disk to the far side *(center)*; consequently, it is turned in the opposite direction. The neutral setting simply disengages the drive wheel from the drive disk *(bottom)*, halting the mower.

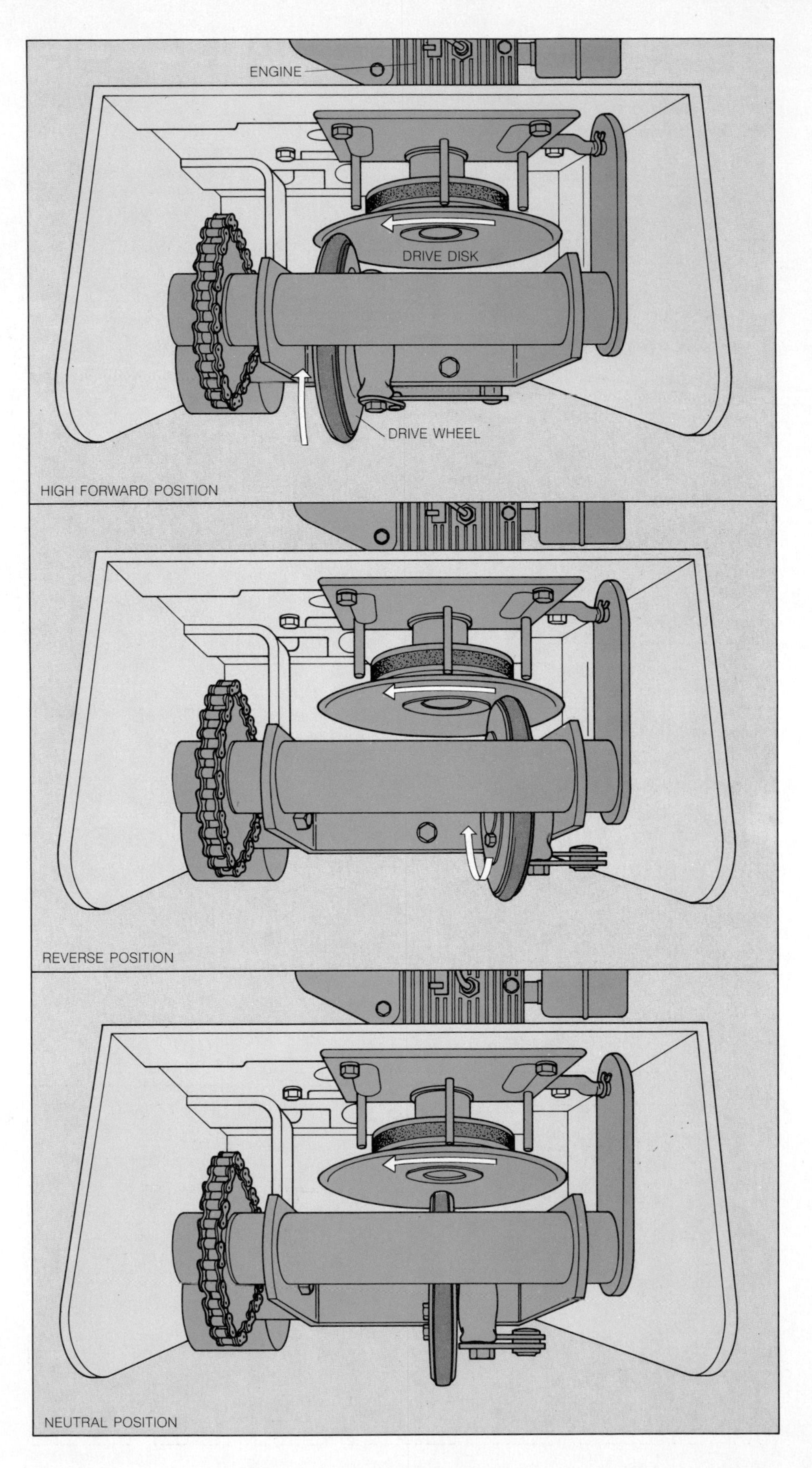

Replacing a Worn Rear Belt

Removing the rear belt. Place the shift lever in neutral, then unbolt and remove the belt guide at the rear of the mower *(below, left)*. Jack up the mower or stand it on end and secure it to a wall. Slip a loop of rope under the hooked end of the idler-pulley spring and lift the spring off its anchor bolt near the rear belt *(below, right)*. Lift the belt off the intermediate, two-pulley spindle and work it away from the rear pulley, sliding it out between the drive disk and drive wheel. Slide the new belt between the drive disk and the drive wheel, raise it and loop it around the rear pulley, and slip it over the two-pulley spindle. Return the idler-pulley spring to its anchor bolt, then lower the mower and replace the belt guide.

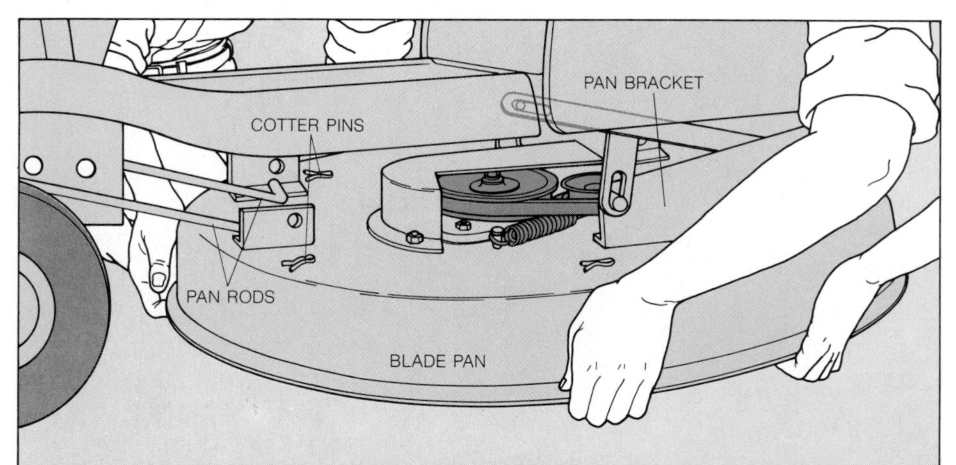

Replacing a Worn Front Belt

1 Lowering the blade pan. Remove the rear belt *(top)* and set it aside. Lower the mower back onto its wheels. Place the blade-control lever in the off position to release tension on the front belt, and remove the cotter pins securing the pan to the pan rods and pan brackets. With a helper, work the pan loose from the rods and brackets, and lower it gently to the ground; take care not to stretch or tear the front belt, which is still attached to the intermediate and blade pulleys.

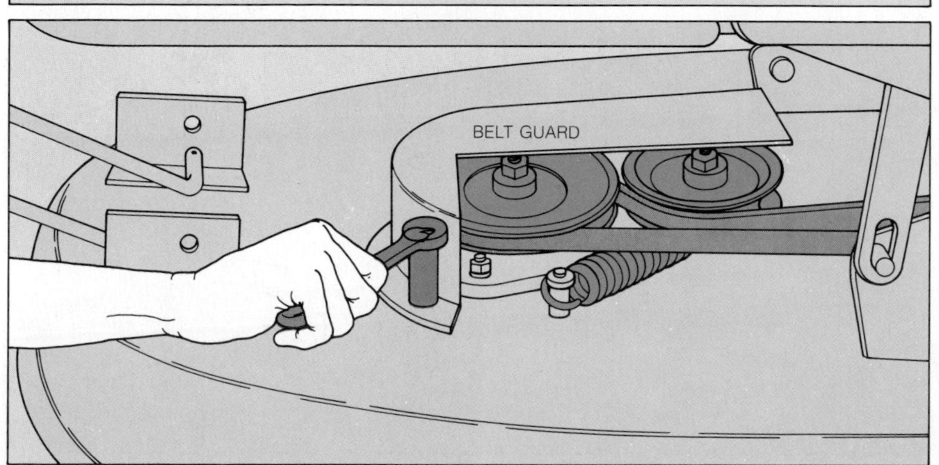

2 Removing the front belt. Unbolt and remove the belt guard. Detach the front idler-pulley spring as you did the rear one *(top right)*. Slide the belt off the front pulley and the two-pulley spindle.

Install a new belt by looping it over the front pulley and the two-pulley spindle, then attach the front idler-pulley spring and replace the belt guard. Raise the blade pan into place and refasten it to the pan rods and brackets with the cotter pins. Finally, replace the rear belt.

Keeping the Drive Chains Tight

Three ways to adjust the chain. To adjust the chain on mowers with rear-axle bearing brackets *(top)*, raise or jack up the mower, loosen the bolts on the brackets at both ends of the axle and slide the entire axle up or down to achieve the chain tension specified by your owner's or service manual. On models with a tension roller *(center)*, loosen the nut on the end of the roller and slide the roller up or down in its slot to alter the tightness of the chain. Retighten the nut after making the adjustment. For mowers with an eccentric cylinder—or cam—to keep the chain tight *(bottom)*, loosen the nut behind the cam and rotate it to achieve the correct tension; retighten the nut. Note: Mowers with tension rollers or cams need not be upended or jacked to adjust the chain tension.

Some mowers will use a combination of these methods—one for each of the chains in the series linking the drive wheel to the rear wheels.

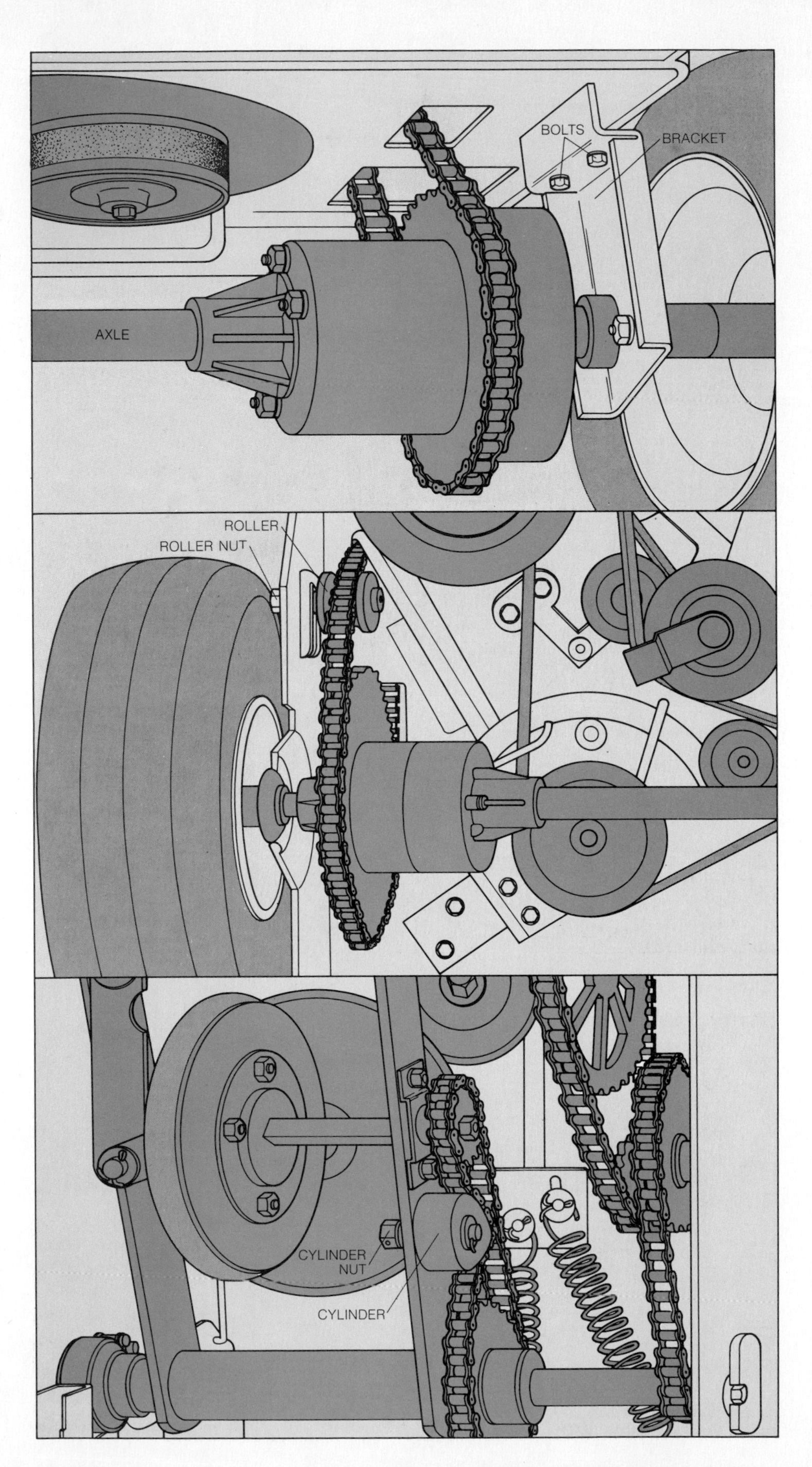

Adjusting the Play of the Drive Wheel

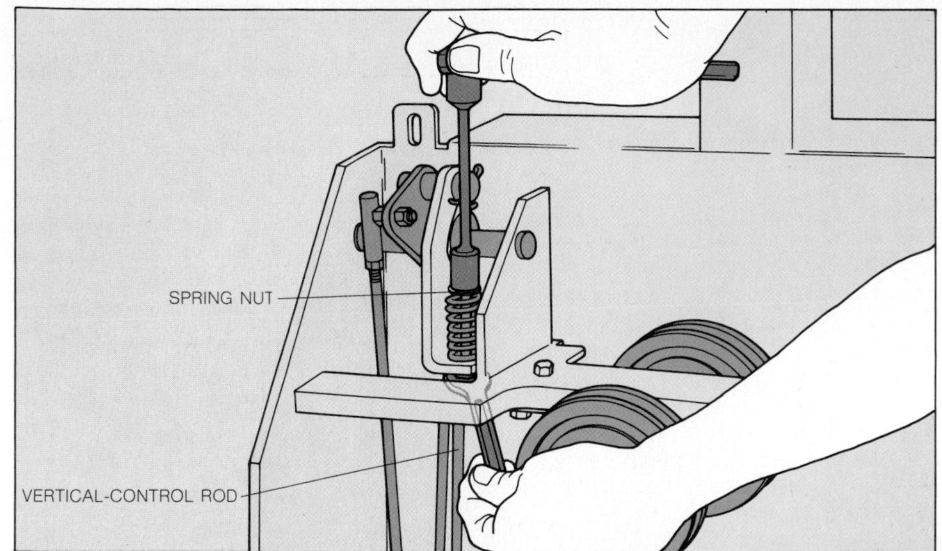

Altering the vertical travel. To adjust the neutral gear setting, remove the blade pan *(page 117)*, place the shift lever in neutral and locate the vertical control rod *(page 115)*, which links the shift lever to the bracket that retains the drive wheel. Using a socket wrench, adjust the spring nut on the control rod while holding the bottom nut secure with a second wrench, to raise or lower the bracket and drive wheel. Position the drive bracket for 1/16 inch of clearance between the drive wheel and drive disk.

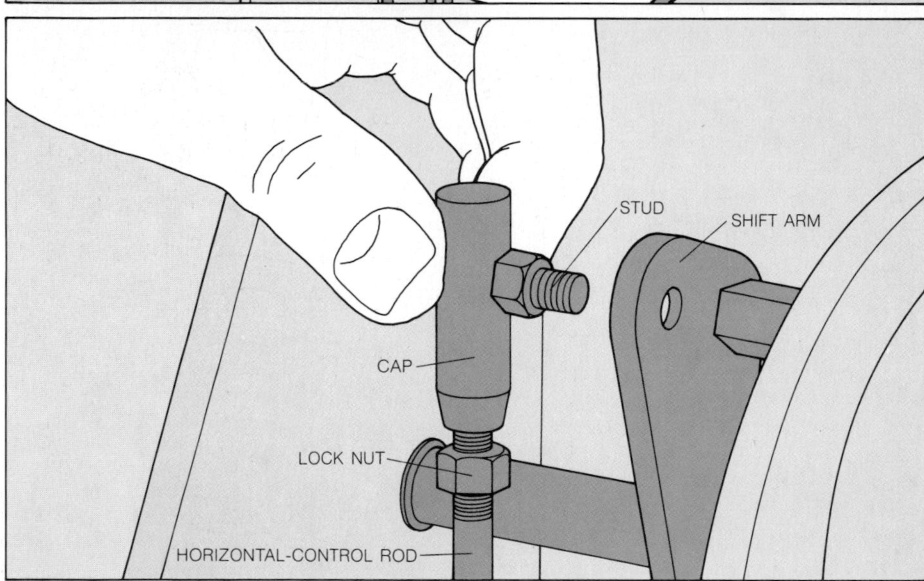

Adjusting the horizontal travel. If the mower slips out of gear when running fast, stand the mower on the tilt bar and move the shift lever into reverse. Locate the horizontal control rod that links the shift lever to the drive wheel *(page 115)*. Remove the nut that holds the threaded stud to the shift arm, and pull the rod away from the arm. Loosen the lock nut below the cap of the rod; screw the cap down a few turns to restrict drive-wheel travel. Tighten the lock nut, and fasten the stud to the shift arm. Move the shift lever to the high-speed forward setting: The drive wheel should travel to the outer rim of the drive disk, stopping just short of the edge.

To adjust reverse gear, disconnect the horizontal control rod, but instead of tightening the cap, loosen it several turns. When the shift lever is moved to reverse, the drive wheel should travel just past the center of the drive disk.

A Brake to Stop the Blade

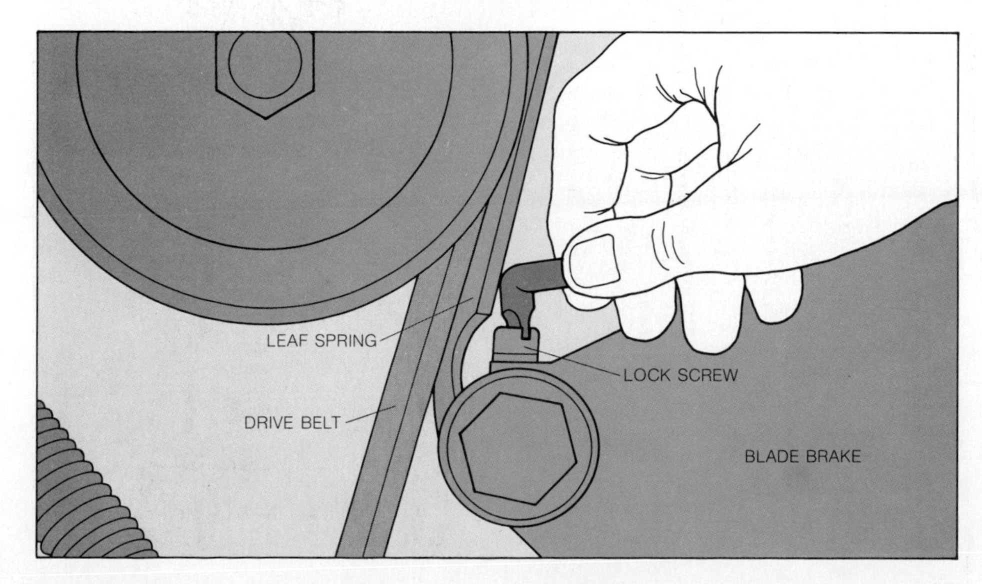

Inspecting and adjusting the blade brake. Periodically have a helper observe the stopping time of the blade after you disengage it: The blade brake should stop the blade in four to six seconds. If it does not, tip the mower up on the tilt bar and secure it to a wall. Using a right-angle screwdriver for easy access, loosen the lock screw on the blade brake. Move the leaf spring closer to the drive belt to stop the blade more quickly, or move it away if the brake grabs too suddenly. Retighten the lock screw.

To test the adjustment while the mower is still upright on the tilt bar, set the blade-control lever to disengage the blade. Grasp the drive belt just below the leaf spring and tug hard; the brake should keep the belt from slipping.

A Whirling String That Trims Weeds

A lightweight weed trimmer with a cutting string that whips around to chop weeds and grass in hard-to-reach places has become a popular laborsaving tool in many gardens. Simple to operate, it is also simple in design: A long handle is fitted with a round plastic case, called a cutting head, that contains a reel of tough nylon line.

Most weed trimmers are powered by two-stroke engines that allow them to be operated at any angle. On some, the engine is at the bottom of the handle, attached directly to the cutting head. On others, such as the one shown below, the engine is mounted at the top of the handle and connected to the cutting head by the drive shaft, which is a flexible cable. The engine twirls the drive shaft, which in turn spins the cutting head.

Aside from the engine, which is maintained and repaired as described in Chapters 2 and 3, a weed trimmer has only a few moving parts; all are easy to care for. The cutting head, which can become clogged with dirt and debris, should be opened and cleaned after every use to keep the cutting line operating smoothly. The drive shaft, because it is in constant contact with its metal housing, needs frequent lubrication—after about every 10 hours of use. (The owner's manual will give specific instructions.)

Normal wear eventually erodes the spool that holds the cutting line and the spindle-like drive gear that turns it. Examine them each time you clean the cutting head, and check to see if any teeth have broken off or become rounded or cracked. If so, replace them with new parts supplied by the manufacturer. New cutting line is available at most hardware stores; buy the type recommended by the manufacturer of your model.

Some weed trimmers, designed with larger-than-average engines and special safety features, are available with optional circular saw-toothed blades to trim brush and small saplings. The blade is cleaned and sharpened like any other metal tool blade.

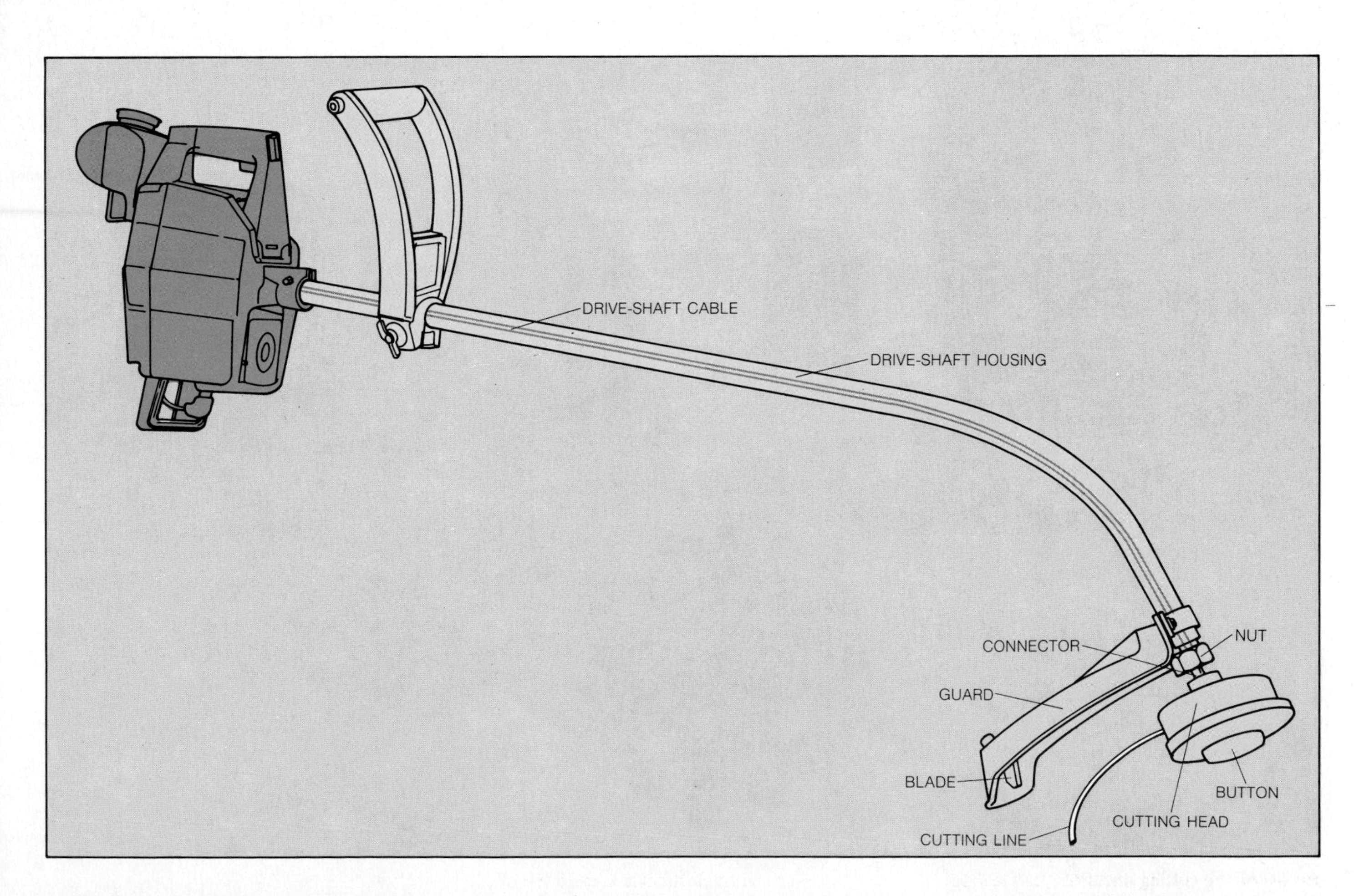

Anatomy of a weed trimmer. The two-stroke engine on this trimmer is mounted at the top of the tool. The drive shaft, a ½-inch flexible steel cable, runs from the engine to the cutting head through a hollow metal housing. At the bottom of the housing, the drive-shaft cable fits through a large nut into a threaded connector. The cutting head, which contains a spool of cutting line, is screwed directly to the connector.

When the engine is running, the drive shaft twirls at high speed, spinning the cutting head. A hard-plastic guard shields the operator from flying debris. When the broad button on the bottom of the cutting head is tapped on the ground, it unreels a length of fresh cutting line; the frayed, used section is sliced off by a small blade mounted on the guard.

Adjustments and Repairs Inside the Cutting Head

Opening the head for inspection. Grasp the nut at the end of the drive-shaft housing with an adjustable wrench, and unscrew the cutting head *(below, left)*. Holding the head in both hands, press the lock tab on its side and turn the retaining ring; the ring will pop off, opening the head *(below, right)*. Wipe out the head; inspect the drive gear and spool. If the teeth of either one are worn or broken, replace the part *(below and page 122)*.

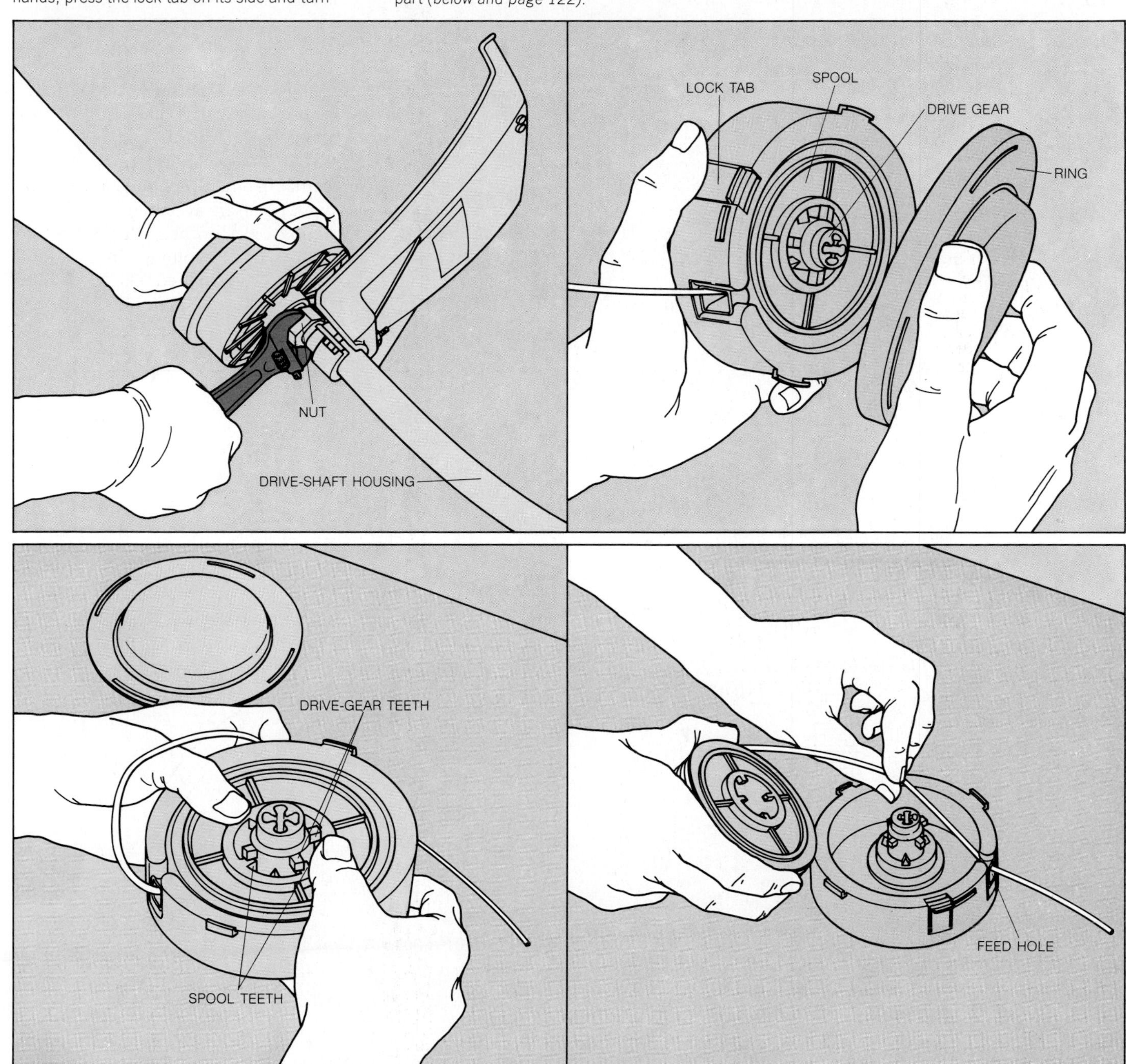

Replacing the cutting line. Press on the spool, twisting it to move its teeth from under the drive-gear teeth *(above, left)*; the spool will pop up. Remove it and pull out any bits of old line.

Wind a new length of cutting line around the spool, working in the direction indicated by an arrow on the spool or as instructed in the owner's manual. Insert the end of the line through the feed hole in the side of the cutting head *(above, right)*, and then fit the spool into place over the drive gear. Press the spool down and twist it to lock it under the drive-gear teeth. Finally, replace the retaining ring.

Replacing the drive gear. Open the cutting head as on page 121, top. Turn the spool and release it just enough to bring its teeth next to those on the drive gear. Turn the spool counterclockwise to loosen the gear. Then remove the spool and unscrew the gear. Twist on a new drive gear and use the spool to tighten it.

If the gear is stuck, and its teeth or the spool teeth are broken or too worn to press against one another, use pliers to twist off the gear.

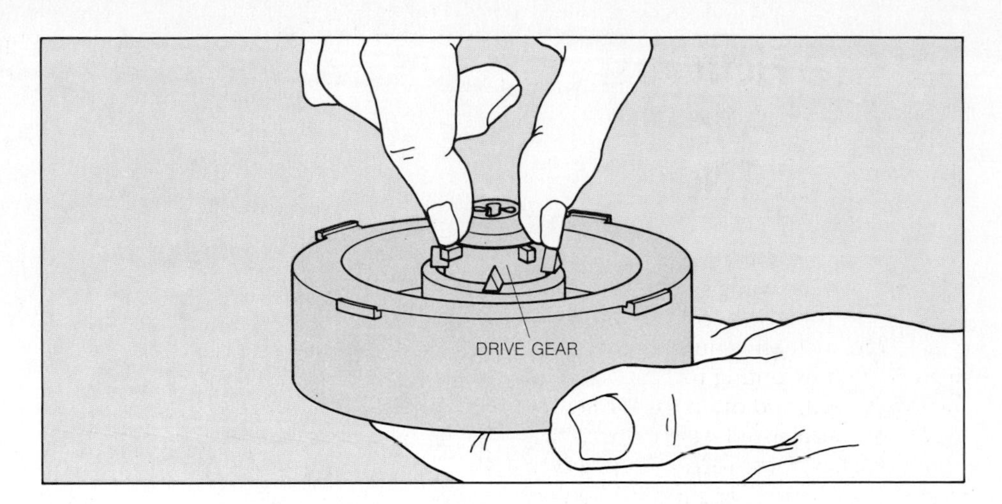

Sharpening the Line Cutter

Filing the cutting blade. Place the weed trimmer on a worktable with the edge of the blade facing up. Sharpen it with a flat file or the flat side of a half-round file. File in one direction only—from the outer to the inner edge of the blade.

If the blade is broken or chipped, you must remove it and replace it with a new one.

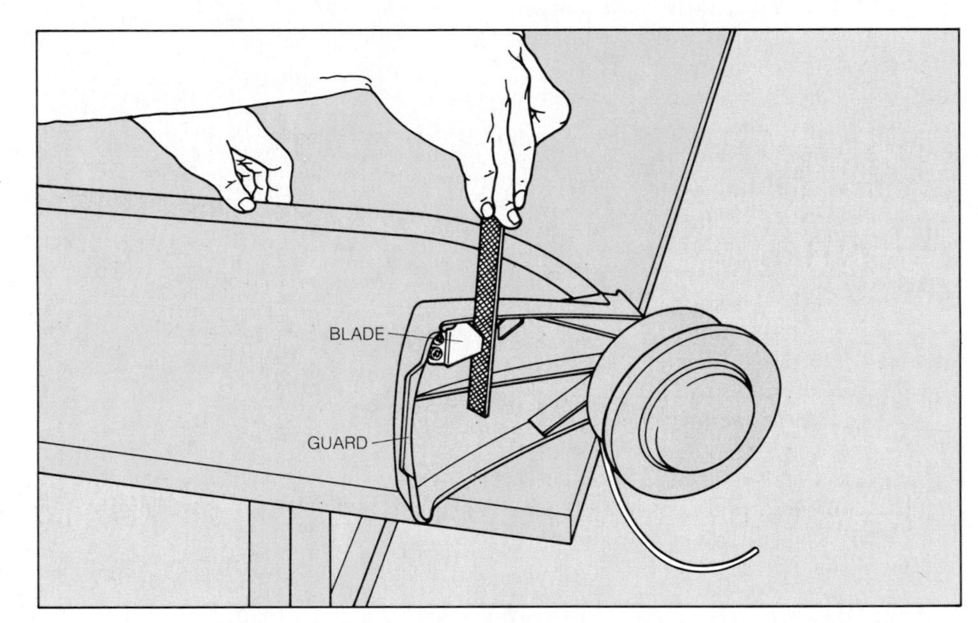

Maintaining the Drive Shaft

Lubricating the cable. Unclamp the drive-shaft housing from the engine casing. Then pull the flexible drive-shaft cable out of the housing. (Note the difference between the two ends of the cable, in order to reassemble it correctly later.) Lay the cable on a clean surface and use a clean cloth to wipe it free of old grease. Then coat the cable with the type of lubricant recommended in the owner's manual, using a cloth to spread it evenly over the surface.

Reinsert the drive shaft into the housing. Fit the top end of the shaft and the housing into the opening in the engine casing, and clamp it in place. Reattach the cutting head.

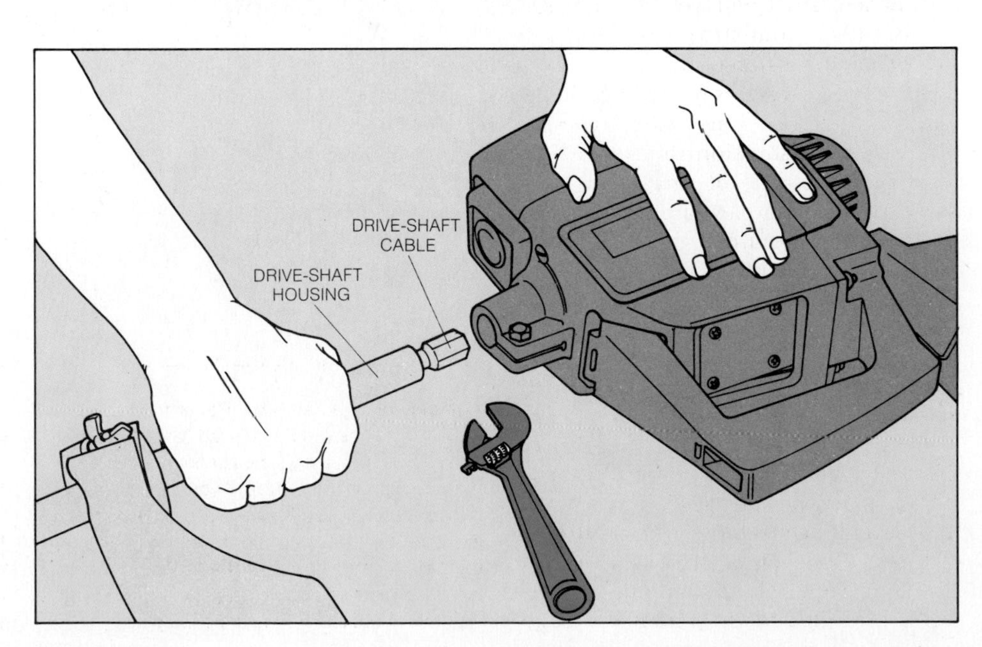

For the Hardest Work of All: A Power Tiller

Every spring the soil-churning blades of garden tillers—crawling spider-like across backyard seedbeds—become a more familiar sight. Increasing numbers of homeowners are now opting to grow some of their own food, and often on a scale that makes power-assisted cultivation a necessity. Luckily, these rotary tillers, driven by sturdy four-stroke engines, are subject to remarkably few mechanical breakdowns. The faults that do crop up are usually in the trouble-prone belt-and-pulley mechanism that turns the tines.

Because the drive belt wears and stretches in use, it must be checked for adjustment every time you use the machine. After every 10 hours of use, you also should examine it for cracks, flaking or other signs of deterioration. If the underside of the belt is shiny, the belt has been slipping on the pulleys and must be tightened. Do not overtighten the belt, however; a belt that is too taut can wear out the machine's bearings prematurely.

When adjusting the belt, check that the pulleys over which the belt rides are aligned, to prevent premature belt wear. Also inspect the transmission pulley and its keyed shaft and setscrew. The setscrew should be tight so that the key, keyway and shaft will not be damaged.

The tiller's four-stroke engine is a slightly larger version of the models used on lawn mowers and is maintained in the same way. Unlike a mower, however, a tiller has a transmission to multiply the force of the engine. If the engine runs and the belt and pulleys are working, but the tines will not turn, it is likely that the transmission is faulty. On the model at right, the transmission is a sealed unit, which can be removed in one piece just by loosening a handful of bolts and replaced by a new transmission from the manufacturer. Other models have more elaborate transmissions; if they break, they must be rebuilt by a repair shop.

During heavy-duty garden work, the tines of a tilling machine often hit stumps or rocks. Like lawn-mower blades *(pages 108-109)*, they should be routinely removed and filed to restore an even edge.

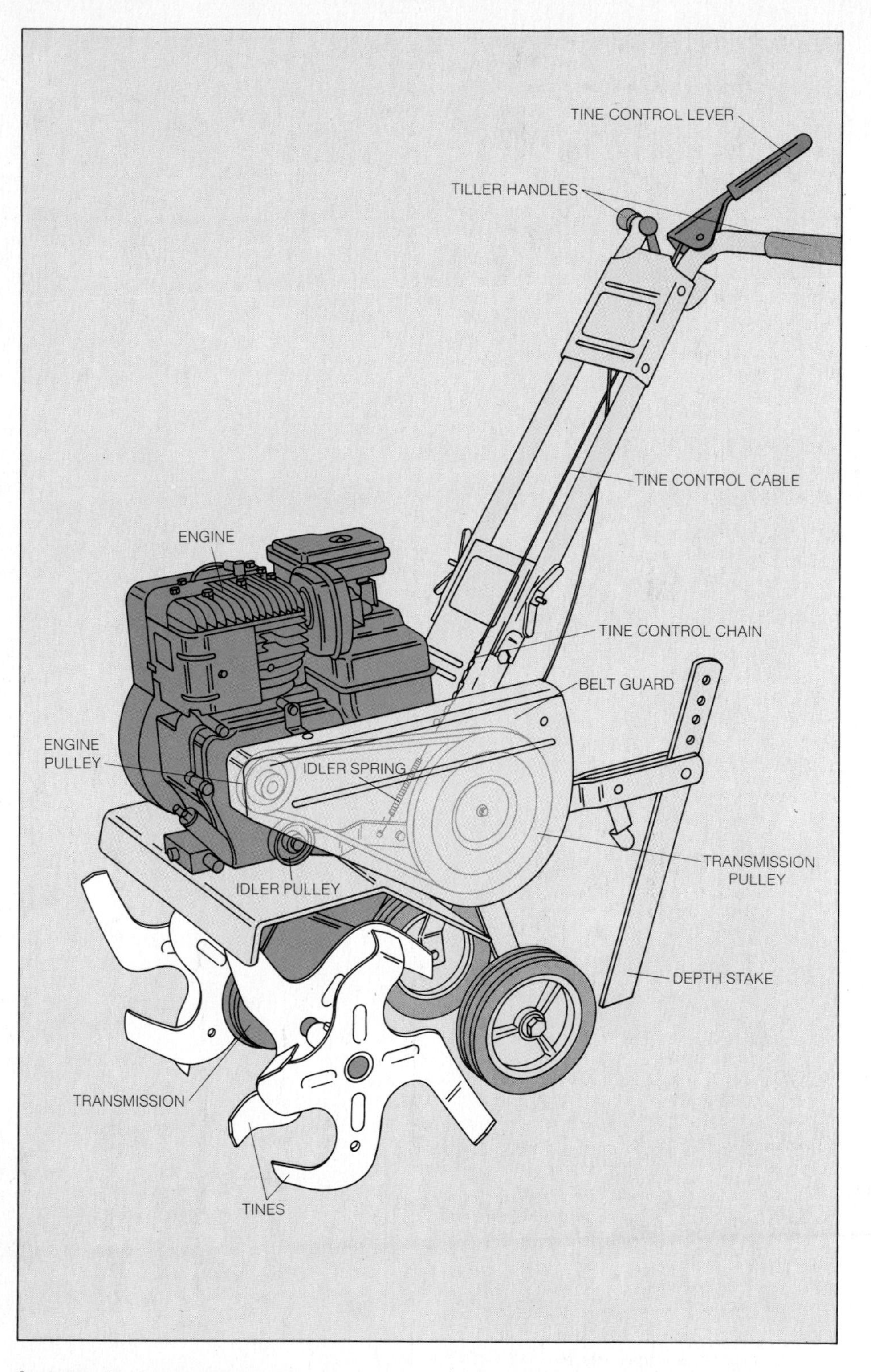

Anatomy of a garden tiller. Power from the four-stroke engine is transferred to the tines by a belt and a set of pulleys, which are engaged by pressure on the tine control lever: When the lever is pushed down, the tine control cable stretches a spring and lifts the idler pulley up against the belt, taking the slack out of the belt and forcing the engine pulley to engage it. The spinning belt in turn rotates the transmission pulley, so that the transmission—a pair of internal sprockets connected by a chain—drives the tines.

For safety, the belt and pulleys are covered by a belt guard. A depth stake behind the tiller can be raised or lowered to regulate the depth of cut and, hence, the speed at which the tiller moves over the ground. The deeper you are tilling, the more slowly the tiller will move.

Adjusting the Belts and Pulleys

Adjusting the tine control cable. Disconnect the spark-plug wire from the spark plug to make sure the engine does not start while you are working. Press the tine control lever and hold the tiller handlebars down to rock the tines off the ground. Pull the starter rope *(below, left)*. If the tines do not rotate, remove the hook at the end of the idler spring from the tine control chain and reinsert it in a higher link *(below, right)*. To make certain that you have not tightened the tine control cable too much, raise the tines off the ground and pull the starter rope without pressing down on the tine control lever. The tines should not rotate. If they do, reposition the idler-spring hook in a lower link of the chain.

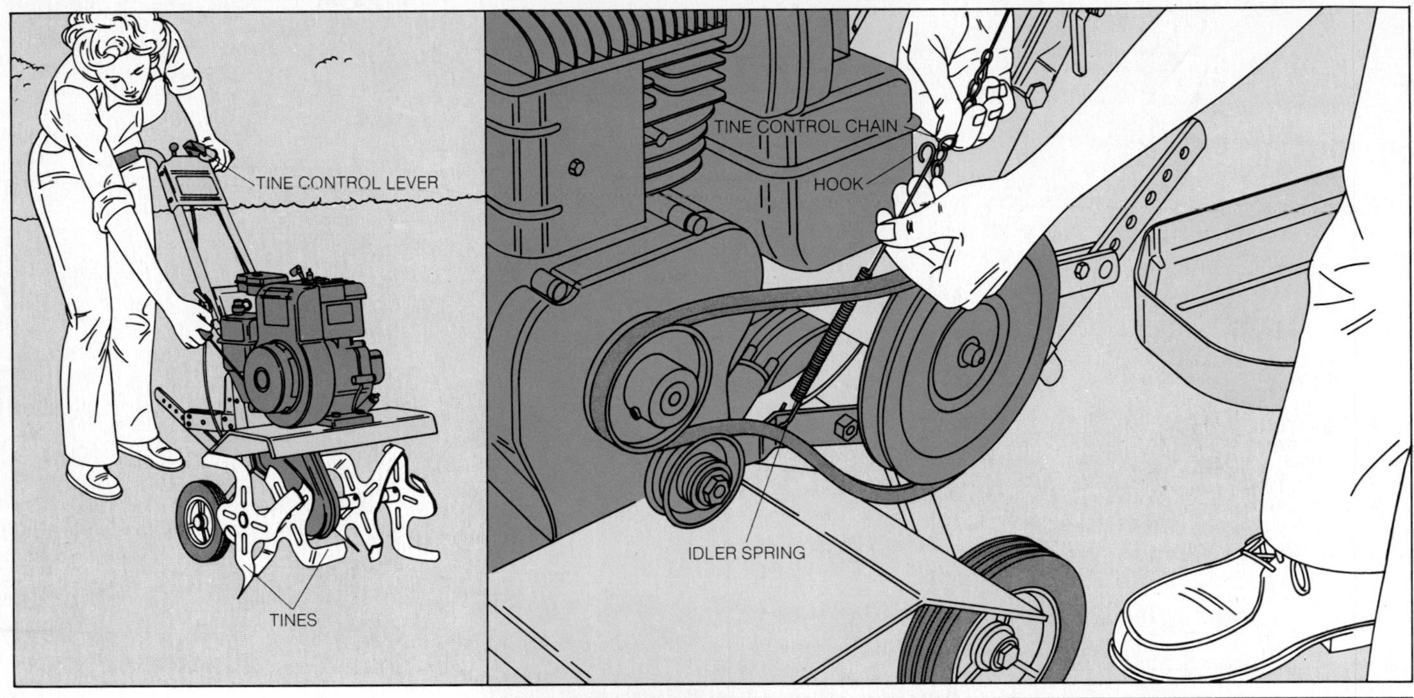

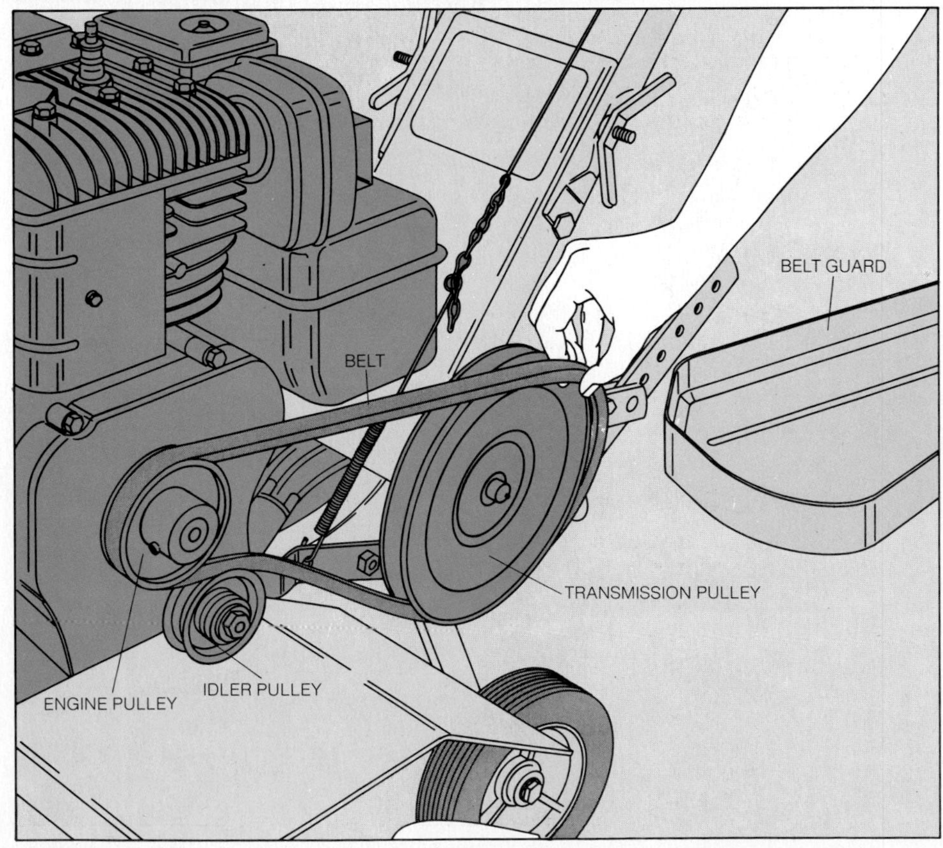

Replacing the belt. Disconnect the spark-plug wire. Remove the bolts that hold the belt guard, and take it off the tiller. Lift the belt off the transmission pulley, then pull it away from the engine pulley and the idler pulley. Wipe off the pulleys with a clean cloth.

To install a new belt, loop it around the engine pulley with the V-shaped side seated firmly in the pulley groove. The flat part of the belt will face the idler pulley. Loop the belt around the transmission pulley. Reinstall the belt guard, and adjust the tine control cable *(above)*.

Aligning the pulleys. Disconnect the spark-plug wire, remove the belt guard and rock the tiller forward onto the tines. Then, pressing down on the tine control lever to engage the belt, sight along the pulley edges from behind the transmission pulley *(below, left)*. If the outer edges of the three pulleys are not in line with each other, use the idler pulley, which is fixed, as a guide to realign the others: Disengage the belt and loosen the engine pulley, using a hex wrench to release the pulley setscrew *(below, right)*. Move the engine pulley back and forth along its shaft until its edge lines up with the edge of the idler pulley. Retighten the setscrew. Align the transmission pulley with the idler pulley in the same way, taking special care to tighten the pulley setscrew so that the key in the pulley-shaft keyway is held firmly *(inset)*.

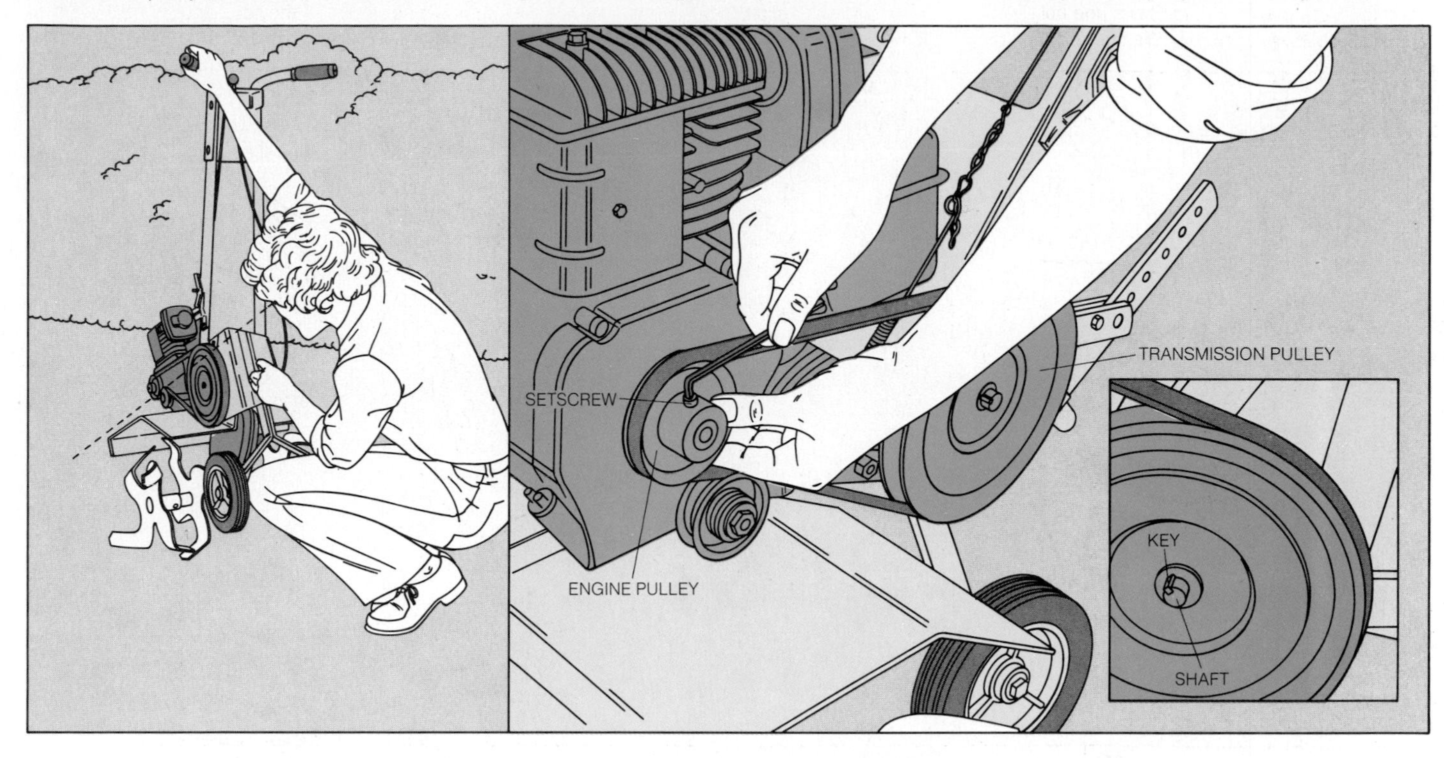

Removing the Tines for Service

Freeing the blades from the shaft. Loosen the two large bolts that secure the depth stake. Remove the stake, and lean the tiller back on its handles. Wash the dirt and rust from the hair-pin clips and retaining pins that secure the tines, and put a few drops of penetrating oil around the pins to help free them. Wearing gloves to protect your hands from the sharp edges of the tines, use pliers to pull out the clips. Pull out the pins; if they are stuck, tap them out with a hammer. Slide the tines off the shaft, noting their positions. Sharpen, straighten or replace any damaged tines. Reinstall the tines in the original positions by sliding them onto the shaft, inserting the retaining pins and attaching the clips.

Picture Credits

The sources for the illustrations in this book are shown below. The drawings were created by Jack Arthur, Roger Essley, Charles Forsythe, William J. Hennessy Jr., John Jones, Dick Lee, John Martinez and Joan McGurren.

Cover: Fil Hunter. 6-9: Fil Hunter. 10, 11: Elsie J. Hennig. 12, 13: Walter Hilmers Studios. 14, 15: William J. Hennessy Jr. from A and W Graphics. 16, 17: Walter Hilmers Studios. 18: Fil Hunter. 21-24: Eduino J. Pereira. 25: Frederic F. Bigio from B-C Graphics. 27-31: Adisai Hemintranont from Sai Graphis. 33: John Massey. 34: Roger Essley. 35-37: John Massey. 39-41: Eduino J. Pereira. 42-45: Adisai Hemintranont from Sai Graphis. 46-53: Frederic F. Bigio from B-C Graphics. 54, 55: William J. Hennessy Jr. from A and W Graphics. 56: Fil Hunter. 58, 59: Graham Sayles. 61-63: Adisai Hemintranont from Sai Graphis. 64-69: William J. Hennessy Jr. from A and W Graphics. 71-80: John Massey. 81-83: Frederic F. Bigio from B-C Graphics. 84-89: Arezou Katoozian from A and W Graphics. 91-93: Elsie J. Hennig. 94: Fil Hunter. 97-105: Frederic F. Bigio from B-C Graphics. 107-113: Elsie J. Hennig. 115-122: Walter Hilmers Studios. 123-125: William J. Hennessy Jr. from A and W Graphics.

Acknowledgments

The index/glossary for this book was prepared by Louise Hedberg. The editors would like to thank the following: Jack Carroll, McCulloch Corporation, Los Angeles, Calif.; Vincent D'Agostino, Tom's McLean Service, McLean, Va.; Chuck Frickey, Briggs and Stratton, Milwaukee, Wis.; Stuart Gannes, Montclair, N.J.; Michael Isser, Mekler/Ansell Associates, Inc., New York, N.Y.; Al Jacobson, Onan Corporation, Minneapolis, Minn.; Lane's Mower Service, Inc., Arlington, Va.; Richard W. Osborn, Homelite Division of Textron, Inc., Charlotte, N.C.; Charles Royster, President, Promower, Inc., Rockville, Md.; Bob Sohl, Deere & Company, Moline, Ill.; Alfred E. Strege, Kohler Company, Kohler, Wis.; Phillip G. Vollmer, Stihl Incorporated, Virginia Beach, Va. The editors would also like to express their appreciation to Robert Cox and Lee Greathouse, writers, for their assistance with the preparation of this volume.

Index/Glossary

Included in this index are definitions of many of the technical terms used in this book. Page references in italics indicate an illustration of the subject mentioned.